NATURAL RESOURCE MANAGEMENT AND POLICY

Series Editors:

David Zilberman
Department of Agricultural and Resource Economics
Berkeley College of Natural Resources
University of California
Berkeley, CA, USA

Renan Goetz
Universitat Girona Depto. d'Economia
Girona, Spain

Alberto Garrido
Department of Agricultural Economics and Social Sciences
Technical University of Madrid (UPM)
Madrid, Spain

For further volumes:
http://www.springer.com/series/6360

EDITORIAL STATEMENT

There is a growing awareness to the role that natural resources, such as water, land, forests and environmental amenities, play in our lives. There are many competing uses for natural resources, and society is challenged to manage them for improving social well-being. Furthermore, there may be dire consequences to natural resources mismanagement. Renewable resources, such as water, land and the environment are linked, and decisions made with regard to one may affect the others. Policy and management of natural resources now require interdisciplinary approaches including natural and social sciences to correctly address our society preferences.

This series provides a collection of works containing most recent findings on economics, management and policy of renewable biological resources, such as water, land, crop protection, sustainable agriculture, technology, and environmental health. It incorporates modern thinking and techniques of economics and management. Books in this series will incor porate knowledge and models of natural phenomena with economics and managerial decision frameworks to assess alternative options for managing natural resources and environment.

The Series Editors

Govinda R. Timilsina • David Zilberman
Editors

The Impacts of Biofuels on the Economy, Environment, and Poverty

A Global Perspective

Springer

Editors
Govinda R. Timilsina
The World Bank
Washington, DC, USA

David Zilberman
Department of Agricultural
and Resource Economics
Berkeley College of Natural Resources
University of California
Berkeley, CA, USA

ISBN 978-1-4939-0517-1 ISBN 978-1-4939-0518-8 (eBook)
DOI 10.1007/978-1-4939-0518-8
Springer New York Heidelberg Dordrecht London

Library of Congress Control Number: 2014935150

Printed on acid-free paper

Springer is part of Springer Science+Business Media (www.springer.com)

Foreword

This is a timely book on a subject of great importance to the world—biofuels and their impact on the economy and development. The production of biofuels lies at the cross section of major challenges for the world. It has the ability to ease some of the world's most challenging energy bottlenecks, but it has implications, positive and negative, for climate change and the environment, and, further, it competes for some of the same resources that are needed for food production and hence have implications for poverty eradication and basic needs. At the crossroads of these important concerns, and with no obvious, black-and-white answers, this is a topic of heated political debate and competing claims. This is an important book that should help in placing the debate on a firmer ground of facts and reason. So whether you are an amateur, wanting to spruce up your vocabulary a little, with terms like lignocellulosic feedstock, microalgae density and photobioreactors, or a specialist policy-maker, looking for cutting-edge information and analysis in the area of biofuels, you should find this an interesting volume.

Biofuels rose to sudden prominence, when, as a response to the oil crisis of the 1970s, several countries attempted to actively promote biofuels to substitute for the use of fossil fuels, especially in transportation. Brazil and the United States introduced national programs for ethanol production around 1979. Biofuel promotional efforts also were undertaken by some other countries, such as China, Kenya, and Zimbabwe, though enthusiasm for these waned when oil prices dropped in the early 1980s.

Subsequent increases in oil prices—this time with alarming price volatility, and a growing concern with climate change and the environment, combined with a long period of stable agricultural commodity prices, induced many countries to consider biofuels a priority once again. Production of biofuels has increased rapidly, by more than 20 % per year since the early 2000s. Government policies, particularly subsidies and blending mandates, have been the key drivers behind the biofuel boom. Today, more than 40 countries the world over, including a number of developing countries from Sub-Saharan Africa and South Asia, have introduced mandates and targets for biofuels. As of today, biofuels account for around 3 % of the total liquid fuels for transportation; if current blending mandates and utilization targets are fully realized, the contribution of biofuels to total liquid fuel consumption for transportation would reach 10 % by 2020.

This rapid, largely policy-driven, increase in biofuel production attracted attention worldwide when agricultural commodity prices, which were moderate and stable for decades, started increasing, culminating in a global food crisis in 2008. Biofuels were pointed to as one of the contributors to this food crisis, as significant amounts of land and other resources for food crops were believed to have been diverted to biofuel production. Meanwhile, questions began to be raised about the climate change and environmental benefits of biofuels, one of the key drivers for its promotion, when increased use of biofuels in the United States and Europe was found to indirectly cause deforestation and loss of biodiversity in developing countries. A long debate thus got ignited on the extent to which biofuels ought to be promoted through costly policies, including million dollars given out annually as subsidies.

These issues attracted the attention of the World Bank, although the Bank has not, to date, financed nor provided technical support to biofuel production. In 2008, the Research Department of the World Bank initiated a comprehensive program to analyze the economic, environmental, and social impacts of significant expansion of biofuels. This edited volume summarizes the key findings of the World Bank study and provides access to a broad knowledge base in the literature on the economics of biofuels. The volume will be of benefit to a wide audience including policy advisors, academic researchers, industry representatives, and members of civil society groups in understanding the economic, environmental, and social implications of biofuels' expansion and the impacts of the policy instruments driving the expansion.

Washington, DC, USA Kaushik Basu

Acknowledgements

This edited volume is a result of series of research undertaken by the Environment and Energy Unit of Development Research Group of the World Bank for the last 5 years on economic, environmental, and social impacts of biofuels. A large number of researchers both from inside and outside the Bank have contributed to the knowledge packaged in this volume. We would like to sincerely appreciate all authors/coauthors who volunteered to write chapters of this volume.

We would like to thank David Roland Holst for his advice and support in preparing the manuscript, Shelley J. Jiang for helping restructuring the initial manuscript of the book and assisting in making difficult technical material more accessible. We owe special thanks to Eunice Kim, Scott Kaplan, Angie Erickson, and B. Elaine Wylie for editorial assistance and support. We always benefitted from discussion with Mike Toman who helped sharpen messages in several chapters.

We wish to acknowledge financial support of the World Bank, especially the Knowledge for Change Program (KCP) Trust Fund, the Berkeley Bio-economy Institute, Cotton Incorporate and Giannini Foundation for supporting research leading to this work, and the Berkeley Bio-economy Conference where early results were presented.

Washington, DC, USA — Govinda R. Timilsina
Berkeley, CA, USA — David Zilberman

Contents

Contributors

Miguel A. Carriquiry is a Professor in the Economics Institute of the Economics Sciences Department at the Universidad de la Republica, in Uruguay. He earned his Ph.D. in Agricultural Economics from Iowa State University. He previously worked as a research scientist at Iowa State University and Columbia University. His current research focuses on the modeling of agricultural commodity markets and the interaction of bioenergy policies and agriculture, with emphasis on land use change.

Jay J. Cheng is a Professor of Bioprocessing and Environmental Engineering in the Biological and Agricultural Engineering Department at North Carolina State University. He holds a Ph.D. in Environmental Engineering from University of Cincinnati. His research interests include brewing, anaerobic digestion of organic waste materials, nutrient recovery from wastewater with aquatic plant, and biomass conversion to biofuel.

Caesar B. Cororaton is a Research Fellow at Virginia Polytechnic Institute and State University. He holds a Ph.D. in economics in Clark University, MA, USA. He has done poverty analyses in several countries including the Philippines, Pakistan, India, Papua New Guinea, Kiribati, and Tonga. He has worked on global and country computable general equilibrium models linked to household survey data.

Stefan Csordás is an economist in the Environmental Policy Program of GIZ in Berlin, Germany. Before joining GIZ, he was an environmental economist at the World Bank in Washington, DC. He holds a Ph.D. in Economics from the University of Basel. His main focus lies on green economy, low-carbon development, and economic and fiscal policies.

Xiaodong Du is an assistant professor in the Agricultural and Applied Economics at the University of Wisconsin–Madison. He received his Ph.D. in economics in 2008 from Iowa State University. His research focuses on economics and policy issues related to traditional and renewable energy resources, technologies, and markets, especially the transportation fuel sector.

Gal Hochman is an Associate Professor in Agricultural, Food, and Resource Economics at Rutgers University. He holds a Ph.D. in Economics from Columbia University in 2004. His current work focuses on the political economy of fuel policy, the economics of petroleum refineries, the economics of renewable energy, energy efficiency, as well as agriculture biotechnology.

Scott Kaplan Department of Agricultural and Resource Economics, University of California at Berkeley

Simon Mevel works for United Nations Economic Commission for Africa as an Economic Affairs Officer. Prior to this appointment, he worked with World Bank Research Department and the International Food Policy Research Institute. He holds a master's degree in International Economics from the University Pau, France. His research interests are economic modeling, international trade, and development economics.

Deepak Rajagopal is an Assistant Professor in the Institute of Environment and Sustainability at the University of California, Los Angeles. He has a Ph.D. in Energy and Resources from U.C. Berkeley. Prior to his current position, he was a researcher at the Energy Biosciences Institute, U.C. Berkeley. His areas of interest include energy and environmental economics, lifecycle assessment, multi-criteria analysis for decision making, and environmental and social impact assessment.

Ashish Shrestha is a researcher at the World Bank's South Asia Sustainable Development Department. He holds an M.A. in Sustainable International Development from Brandeis University. He is currently working on energy, environment, and climate change areas with a focus on rural, renewable, off-grid energy solutions, and the mitigation of short-lived climate pollutants and greenhouse gas emissions.

Jevgenijs Steinbuks is an Economist at the World Bank Development Research Group. Before joining the World Bank, he worked as a research economist at the Center for Global Trade Analysis, Purdue University; the Electricity Policy Research Group, University of Cambridge; and Credit Research Center at Georgetown University. He holds his Ph.D. in Economics from the George Washington University. His current research focuses on sustainable resource and land use, electric power, and energy demand and efficiency.

Govinda R. Timilsina is a Senior Economist at Development Research Group of the World Bank. Prior to this appointment, he was a Senior Research Director at Canadian Energy Research Institute in Calgary, Canada. He holds a Ph.D. in energy economics from the Asian Institute of Technology, Bangkok, Thailand. He has 20 years of experience across a broad range of energy and climate change at national and international levels. His current research focuses on biofuels and renewable energy, low carbon economic development, and green growth.

David Zilberman has been a professor in the Agricultural and Resource Economics Department since 1979. He holds a Ph.D. in Agricultural and Resource Economics from U.C. Berkeley. He is a fellow of the American Agricultural Economics Association and the Association of Environmental and Resource Economists. His research interests are in agricultural and nutritional policy, economics of technological change, economics of natural resources, and microeconomic theory.

List of Figures

List of Tables

Chapter 1
An Overview of Global Markets and Policies

Govinda R. Timilsina and Ashish Shrestha

1.1 Background

Following the oil crisis of the 1970s, countries looked to biofuels to substitute the use of fossil fuel in transportation. Brazil and the United States (US) governments impelled national programs for ethanol production (Worldwatch 2007) around 1979; meanwhile, some countries (e.g. China, Kenya, and Zimbabwe) acted in response to the oil crisis but were not able to sustain biofuel production (Liu 2005; Karekezi et al. 2004). When oil prices decreased again, the impetus for alternative fuels retreated—except in Brazil. Current drivers of the alternative energy supply include issues of energy supply security, oil price volatility, climate change, production costs, and more. Subsidy is the main policy instrument to incentivize production, although production costs are dropping.

However, concerns about the sustainability of biofuel feedstock production, in particular, the impacts on food supply, the land use change associated with it and the resulting greenhouse gas (GHG) emissions have mitigated some of the enthusiasm for biofuels in recent years and may affect future demand. Controversies regarding the scaling up of biofuel production gained prominence with rising food prices and the consequent global food crisis in 2007–2008. With significant amounts of food crops being diverted to biofuel production, such as in the United States, where ethanol production consumes about 10 % of annual global corn production, the role of biofuel production on food security has drawn additional scrutiny (REN21 2013). Biofuel was expected to help reduce GHG emissions due to the sheer size of the transportation sector's energy consumption in most economies, yet the conversion

G.R. Timilsina (✉)
The World Bank, 1818 H Street, N.W., Washington, DC 20433, USA
e-mail: gtimilsina@worldbank.org

A. Shrestha
The World Bank, Yak & Yeti Hotel Complex, P.O. Box 798, Kathmandu, Nepal
e-mail: ashrestha1@worldbank.org

G.R. Timilsina and D. Zilberman (eds.), *The Impacts of Biofuels on the Economy, Environment, and Poverty*, Natural Resource Management and Policy 41,
DOI 10.1007/978-1-4939-0518-8_1,

of forest lands and pastures for the cultivation of biofuel feedstock could release more GHGs than biofuels reduce through substitution of petroleum.

Based on its type of input (feedstock) or technology used to convert the feedstock into fuel, biofuel typically has been classified into two generations. First generation biofuels utilize plants' sugar or starch (e.g., sugarcane, sugar beet, cereals, cassava) for ethanol or oilseed (e.g., rapeseed, sunflower, soybean, palm oil) for biodiesel (OECD/FAO 2008). First generation biofuels directly compete with food supply and have been produced at commercial levels for many years. On the other hand, production of second generation biofuels can coexist with food production because it can utilize feedstocks that do not compete with food supply (e.g. jatropha, micro-algae) or use advanced technologies to convert lignocellulosic biomass (e.g. agricultural and forest residues). Production of biofuel from cellulosic biomass enables the utilization of 100 % of the plant parts (including agricultural residue such as corn husks), although these feedstocks are more expensive to convert to energy. As cellulosic biomass is the most abundant biological material on earth, second generation biofuels could even expand its feedstock variety if successfully scaled to commercial production (OECD/FAO 2008). Converting micro-algae to biodiesel appears most promising, since it yields 80 % or more of its dry weight as oil, whereas some other feedstock yield only 5 % of their dry weight (Chisti 2008). Micro-algae is also a resilient plant that can grow in polluted aquifers or salt water and thus does not apply pressure on demand for arable land.

1.2 Production, Consumption, and Trade

World production of fuel ethanol has grown at an average rate of 14 % per year, between 2004 and 2012, although production leveled off in 2011 for the first time since 2000 and, in fact, decreased in 2012 by about 1.3 % by volume from 2011. Most of this reduction in production originated in the US, partly due to high corn prices that resulted from the mid-year drought (REN21 2013). The US and Brazil together accounted for almost 87 % of the 83.1 billion liters produced globally in 2012. Global production of ethanol annually from 2004 to 2012 is shown in Fig. 1.1. Brazil led in ethanol production until 2006, when the US reached over 18 billion liters by a 20 % increase from the previous year (REN21 2008). Since then, the US has been the dominant producer of ethanol by a considerable margin. Other recent leaders in ethanol production include France, China, and Canada, while Germany, Spain, Colombia, Thailand, Belgium, and India are also engaged in commercial production of ethanol.

Compared to ethanol, the aggregate production of biodiesel is much lower but is growing at a higher rate and continued to expand even in 2011 and 2012, growing by more than 18 % total from 2010 to 2012, as ethanol production contracted. Biodiesel production averaged slightly greater than 35 % growth per annum between 2004 (2.3 billion liters) and 2011 (22.5 billion liters). Traditionally, biodiesel has been championed in the European Union (EU), where Germany, France, and Italy

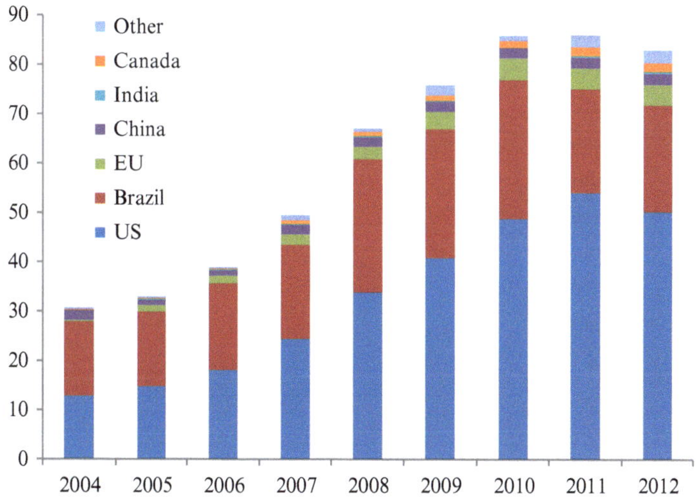

Fig. 1.1 World ethanol production (billions of liters). *Source*: REN21 (2005, 2006, 2008, 2009, 2010, 2011, 2012, 2013), RFA (2008)

led with 90 % of the world's production until 2004 (OECD/FAO 2008). By 2007, however, the EU contributed less than 60 % of biodiesel as the US surpassed French production to become the second biggest producer after Germany (F.O. Licht 2008). Around this same time, the EU began to outsource biodiesel processing to countries such as Indonesia, Malaysia, and Argentina (OECD/FAO 2008). US biodiesel production increased dramatically (by 159 %) to almost 3.2 billion liters in 2011, making the US the leading individual producer of biodiesel, as a government mandate required refiners to blend 3.1 billion liters of biodiesel with diesel fuel in 2011 or else be levied steep penalties (Stebbins 2011). In 2012, Argentina also surpassed Germany in biodiesel production to claim second place, leaving the EU to contribute just over 40 % of world biodiesel production. Although US biodiesel production in 2012 (3.6 billion liters) was up only slightly over 2011 levels, it is approaching the target set by the Environmental Protection Agency (EPA) under the federal Renewable Fuels Standard (RFS), which requires 4.8 billion liters (1.28 billion gallons) of biodiesel to be blended in diesel fuel in 2013 (US EPA 2012) (Fig. 1.2).

Overall, while the US and EU continue to dominate production of ethanol and biodiesel, respectively, production of both biofuels is growing rapidly in Asia and more slowly in Africa as more feedstock becomes available (REN21 2013). Growth is also anticipated in the production of advanced biofuels from lignocellulosic feedstock worldwide, albeit still on a relatively modest scale. US production of such biofuels reached 2 million liters in 2012 and is expected to reach 36 million liters in 2013, partly due to demand from the armed forces (REN21 2013). For example, the US Navy signed contracts to purchase around 1.7 million liters of advanced biofuels in December of 2011 and has pledged to use 50 % fossil fuel alternatives, amounting to 2.3 billion liters of biofuels annually, by 2020 (Chicon 2011).

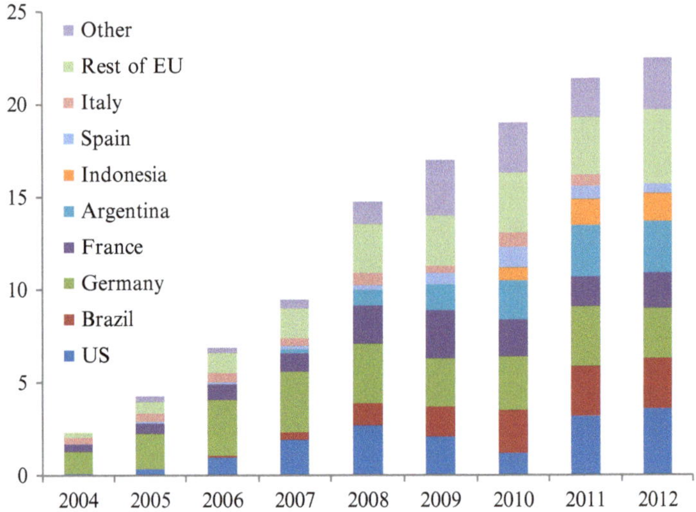

Fig. 1.2 World biodiesel production (billions of liters). *Source*: EBB (2013); EIA (2009); REN21 (2005, 2006, 2008, 2009, 2010, 2011, 2012, 2013)

In China, around 3 million liters of ethanol from corn cobs were produced in 2012, and while Europe also boasts several operational advanced biofuel plants, each has only managed to produce small volumes thus far (Chicon 2011).

The consumption of biofuels has been modest in contrast to the rate of growing production worldwide. In the transportation sector, biofuel demand was above 2 % in only three countries (IEA 2006)—Brazil, Cuba, and Sweden—by 2004, and world transport consumption of biofuel remained about 3 % of the global gasoline consumption of 1,330 billion liters in 2011 (REN21 2012). By 2012, liquid biofuels accounted for an estimated 3.4 % of global road transport fuels (IEA 2013a), as well as a very small but increasing share of aviation and marine fuels, and represent the largest share of transport fuels derived from renewable energy sources (IEA 2011). In some countries, the share of biofuels in road transportation is already considerably higher; for example, 20.1 % in Brazil, 4.4 % in the US and 4.2 % in the EU as of 2010 (IEA 2013b). In order to reduce fuel costs and GHG emissions, airlines around the world are showing greater interest in aviation biofuels, and several of them, including Aeromexico, Finnair, KLM Royal Dutch Airlines, Lufthansa, Thai Airways, United Airlines, and Alaska Airlines started to run commercial flights with various biofuel blends in 2011(REN21 2012).

Based on the increment differential between potential for technical production of biofuels and expected domestic transport energy demand, few countries other than Brazil have export capabilities, but buoyed by subsidies and combined with higher prices for Brazilian ethanol due to poor global sugarcane harvest, the United

Table 1.1 World biofuel trade in 2011 (millions of liters)

Fuel Ethanol			Biodiesel		
Exporter	Importer	Volume	Exporter	Importer	Volume
Brazil	US	325	Argentina	EU-27	1,611
Canada	US	36	Canada	US	103
El Salvador	US	46	EU-27	EU-27	4,812
Jamaica	US	109	EU-27	Norway	34
Trinidad & Tobago	US	225	EU-27	US	40
Brazil	EU-27	49	Indonesia	EU-27	1,225
Egypt	EU-27	28	Norway	EU-27	96
Guatemala	EU-27	17	US	EU-27	133
Pakistan	EU-27	23	US	Norway	26
Peru	EU-27	19	US	Canada	10
Russia	EU-27	12	US	Taiwan	28
US	EU-27	18	US	Israel	10
US	Brazil	1,500	US	Malaysia	8
EU-27	EU-27	1,572	US	Australia	6
			US	India	50

Source: REN21 (2012), Cooper (2012)
Note: This is not an exhaustive listing of biofuel trade in 2011. Indicative traded volumes only, not including significant overall export/import figures. EU-27 to EU-27 indicates trade within the European Union

States, which was a net biofuel importer until 2010, saw its exports rise nearly threefold from 1.5 billion liters in 2010 to 4.5 billion liters in 2011 (Cooper 2012). About one-third of US exports flowed to Brazil, where ethanol production was down by almost 18–21 billion liters in 2011 relative to about 25.5 billion liters in 2010 as declining investment in new sugarcane assets and plantations since the 2008 financial crisis, high world sugar prices, and poor sugarcane harvests due to unfavorable weather all took their toll (REN21 2012). Brazil, which was the world's leading ethanol exporter for many years continued to lose international market share to the United States, especially in its traditional markets in Europe (REN21 2012). Since the EU targets biofuels for domestic consumption and energy diversification, it remains the major importer of biofuels. Meanwhile, Argentina and Indonesia, two countries with significant differentials between production and demand, have emerged as the main exporters of biodiesel to the EU, exporting over 1.6 billion liters and over 1.2 billion liters, respectively, although this is still exceeded by trade in biodiesel within the EU (see Table 1.1).

Trade opportunities are further distorted by sustainability regulations (e.g. EU), bans on imports (e.g. Thailand), and tariffs (e.g. India) or subsidies (e.g. OECD) among governments protecting domestic agricultural and biofuel industries. For example, a blender's tax incentive (no longer available) encouraged the import of ethanol into the US for the purpose of blending and re-exporting. Biofuel trade is expected to increase in the long term due to countries' biofuel targets against inherent comparative advantages, and South America, Central America, and Africa show export potential based on the differential between production potential and demand (Doornbosch and Steenblik 2007). Tropical countries (such as Malaysia, Indonesia, Brazil, and the Philippines) have 2–3 times higher productivity for biofuel feedstocks under normal water conditions (Girard and Fallot 2006).

1.3 Pricing

The average world ethanol increased steadily from around US$0.41/L in 2006 to approximately US$0.85/L in 2012 (US$1.20/L gasoline equivalent) but has come down to US$0.62 in 2013 (OECD/FAO 2012, 2013). Similarly, the average world price for biodiesel in 2012 of about US$1.55/L of gasoline equivalent was higher than in the previous 5 years, when prices ranged between US$0.90 and US$1.50 per liter, but it has come down to US$1.51 in 2013 (OECD/FAO 2013).

Figure 1.3a displays average annual producer price for ethanol in three of the major ethanol producing areas: the US, EU, and Brazil. Whereas the price of ethanol in the EU, which produces significantly less ethanol than the US or Brazil, exhibits a relatively steady upward trend, ethanol prices in the US and Brazil are subject to the immense volatility in the commodities markets for the feedstocks of choice in these countries, corn and sugarcane, respectively. The US domestic price had dropped from about US$0.60/L in 2011 to as low as US$0.55/L in 2012 until

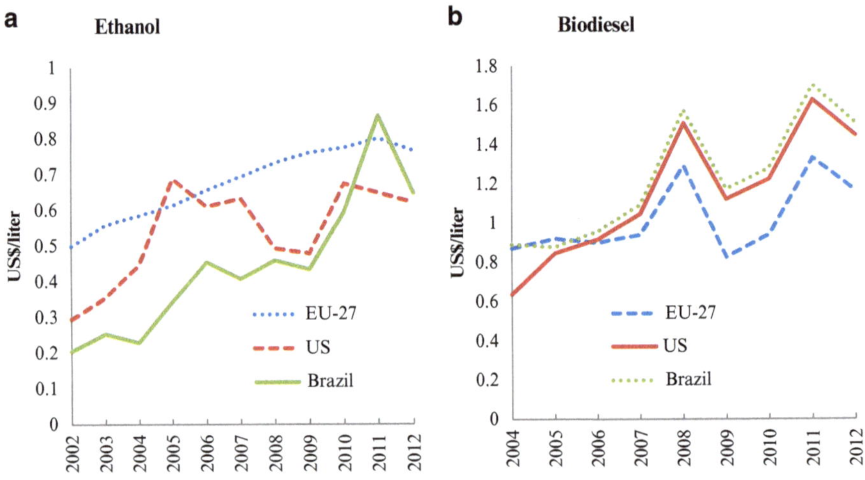

Fig. 1.3 Biofuel prices (US$/L). *Source*: OECD-FAO (2013). *Note*: Prices converted to US$/L using currency exchange rates from World Bank (2013)

the mid-year drought raised it back to 2011 levels (EIA 2013). Brazil, which has traditionally been the lowest cost producer of ethanol, saw its producer price for ethanol rise drastically between 2009 to 2011on account of poor sugarcane harvests due to unfavorable weather and high world sugar prices. Although the price has since come back down, it remains significantly above historical levels and still higher than the producer price for corn-based ethanol in the US. Figure 1.3b exhibits the average annual producer price for biodiesel in the US, EU, and Brazil. While price fluctuations for biodiesel in these regions have moved largely in tandem since 2007, the EU can be seen to feature the lowest price by a considerable margin, with producer prices in the US and Brazil relatively similar. It bears noting that both the US and Brazil are overwhelmingly reliant on soybean oil, another popular commodity, for feedstock, whereas European production of biodiesel is characterized by more diversified portfolio of feedstock such as rapeseed.

1.4 Policies

Whether in the interest of boosting agricultural production or energy security or GHG mitigation, biofuel programs have continued to escalate in recent years. As of early 2013, policies promoting the use of renewable fuels in the transport sector have been identified at the national level in 49 countries, up from 46 identified the year before (REN21 2013). Many nations already regulate biofuel production by mandates on transport fuel composition, and new national blend mandates were introduced in 2012 in South Africa, Turkey, and Zimbabwe (REN21 2013). Brazil, which maintains a variable ethanol blend mandate, reduced the mandated blend level from 24 to 18–20 % in 2011, partly in response to poor sugarcane yields in recent years (REN21 2012). The EU Biofuels Directive of 2003 aimed for 5.75 % biofuel mix in transport fuel consumption by 2010 and 10 % by 2020 (USAID 2009). The original Renewable Fuel Standard program (RFS1) in the United States, created under the Energy Policy Act of 2005, required 7.5 billion gallons (over 28 billion liters) of renewable fuel to be blended into gasoline by 2012 (US EPA 2013). Under the Energy Independence and Security Act (EISA) of 2007, the RFS program was expanded to include diesel and increased mandated volume of renewable fuel blending to 36 billion gallons (over 136 billion liters) by 2022 (US DOE 2008). This latest iteration of the Renewable Fuel Standard program (RFS2) also established separate categories of renewable fuel (e.g., cellulosic) and set separate volume requirements for each category. Other countries like India, Indonesia, Malaysia, and Vietnam have set combinations of blended fuel targets, fiscal incentives, and tariffs to boost their domestic production or to export to the EU (USAID 2009). Biofuels blending mandate and targets are provided in the Appendix. Besides direct endorsements, governments have affected market pricing via national tariffs or subsidies in order to aid developing biofuel industries. The Brazilian government, for example, had initially set market prices, offered subsidies, and guaranteed loans in order to establish the major biofuel industry that exists independently today. Due to recent declines in ethanol production, it has announced financing for agribusiness to increase sugarcane yields, along with loans of US$2.6 billion to sugar companies (REN21 2012). In addition, nearly US$1 billion in loans and grants through 2014 has been committed to support the development of new techniques for creating fuel from sugarcane bagasse, doubling an earlier pledge (Nielsen 2012). In the United States, the Farm Bill of 2008 instated a tax credit for cellulosic ethanol of US$1.01 per gallon (US$0.27/L), a tax credit that has been extended to the end of 2013. The US$1/gal (US$0.26/L) biodiesel tax credit has also been reintroduced for 2013 and applies retroactively to the end of 2011 when that legislation expired (US DOE 2013). Other recent fiscal support for biofuels has been initiated in Australia, where US$15.7 million in grants was pledged for the development of advanced biofuels (IEA/IRENA 2013).

However, biofuel support policies in Europe and the United States remain under review, and pressure continues to mount, especially on support for first-generation fuels, due to concerns about their potential impacts on food production and on land, biodiversity, and water, as well as net GHG emissions from biofuels life-cycle processes (REN21 2013). The European Commission proposed limiting the share of first-generation biofuel in total transport fuels to 5 %. It also plans to phase out subsidies for food crop-based biofuels production by 2020. If enacted, this cap along with the proposed fourfold counting of the shares of advanced biofuels, would effectively mandate the remaining 5 % required to meet the EU's 10 % biofuels target to be realized through second-generation biofuels (Lane 2012). In the United States, although the Renewable Fuels Standard remains in place, the cellulosic fuel target was reduced for the second year in a row, curtailed from 1.9 billion liters (500 million gallons) to 39.7 million liters (10.5 million gallons) (Herndon 2012).

1.5 Investments in Capacity

Whereas investments in biofuel refineries in 2008 were estimated at US$15–16 billion (including venture capital investment of more than US$350 million for cellulosic ethanol alone), biofuel investments had dropped to US$6.8 billion in 2011, down 20 % from the previous year, and to US$5 billion in 2012 (REN21 2009, 2012, 2013). About three-fourths of this investment was made in developed countries. Of the approximately 650 ethanol plants around the world (with a total annual capacity of about 100 billion liters), many are operating below nameplate capacity, while others have closed because of fluctuating demand and reservations regarding the environmental sustainability of first-generation ethanol (REN21 2013). At the beginning of 2012, the US counted 209 ethanol plants in operation with total nameplate annual capacity of over 56 billion liters, representing an increase in 5.3 billion liters of ethanol relative to the previous January (RFA 2013). While two more ethanol plants came online in the US in 2012, production capacity actually declined by more than 700 million liters due to some temporary closures, although four more ethanol plants with a combined capacity of almost 600 million liters are under construction in 2013 (RFA 2013). Brazil had a total of 440 plants with the capacity to produce 37 billion liters in 2011–2012, but excess cane milling capacity in Brazil meant that ethanol production could be increased by another 30 % using existing milling capacity (REN21 2012). Brazil's refining capacity is expected to expand further as new plants come into operation, but new investment has been relatively low in 2009–2011 compared to previous years (REN21 2012). It is worth noting that the size and capacity of plants differ with respect to the type of feedstock that they would process. For example, the US primarily uses corn, which has a longer storage life; therefore US plants have about three times the average capacity of Brazilian plants that treat sugarcane. Elsewhere, new ethanol plants continue to be launched, such as the 54 million liter/year Green Future Innovation Inc. plant that initiated production in the Philippines in January 2013 (REN21 2013).

As demand for biodiesel continues to increase, investments in production capacity can be seen in new plants opening around the world. For example, Cargill (USA) commissioned its first soybean oil based biodiesel plant in Brazil in 2012, and Lignol Energy (Canada) invested US$1.2 million to restart a 150 million liter/year biodiesel plant in Darwin, Australia this year (REN21 2013). Biodiesel investments in the EU resulted in steady increases in production capacity from 24.9 billion liters in 2010 to 25.1 billion liters in 2011, with 22 % of this capacity located in Germany and 20 % in Spain, to over 26.6 billion liters in 2012, although some of this capacity should be considered as idle capacity since some plants have not been in operation for several years and would not be able to resume production this year (EBB 2013). Biodiesel production capacity in the US stood at 11 billion liters from 190 plants in 2011, with 14 new plants estimated at a total of 1.5 billion liters of capacity under construction at the year's end (REN21 2012). Production capacity also expanded significantly in Argentina to 3.8 billion liters in 2011, up almost 36 % from 2010 levels, and even though it produced less biodiesel than Argentina in 2011, Brazil developed far more production capacity, reaching over 6.5 billion liters from 70 plants by the end of 2011 (USDA 2011). Regarding feedstock, biodiesel producing countries have identified local feedstock for cost-effective substitution in producing biodiesel, such as soybean oil in Brazil and Argentina, and rapeseed oil in Canada, and investments have been made to research the regional potential for processing jatropha in Africa, India, and China (Keeney and Nanninga 2008).

Advanced biofuels are generating interest, and investment as well, mainly in the US and Europe, but production volumes remain at pilot scale, although several companies claim to be close to commercial production (REN21 2013). Since 2003, a European company that was founded by DaimlerChrysler, Dutch Shell, and Volkswagen has been researching the conversion of wood waste through the Fischer–Tropsch pathway (REN21 2006). Eighty advanced biofuel companies in the US, 30 of which are located in California, are already producing small volumes, and two advanced biofuels demonstration plants using lignocellulosic biomass and algae were being expanded to near-commercial scale as of early 2013 in Australia (REN21 2013). However, development of advanced biofuels has encountered some recent setbacks as well. The US Court of Appeals ruled that the Environmental Protection Agency must revise its cellulosic ethanol volume projections for 2012, raising doubts about the 2013 standard, although the mandate for the broader category of advanced biofuels was not addressed by this ruling (REN21 2013). Even as one of the early advanced biofuel companies, IOGEN Energy Corporation (Canada), and its current owner Shell Oil abandoned plans to develop a commercial-scale cellulosic ethanol plant in Manitoba, Canada, the aviation industry, dependent on the uncertain long-term supply of petroleum fuels and facing a lack of suitable alternatives, has emerged as an important stakeholder in the development of advanced biofuels: Boeing, Airbus, and Embraer were working together on biofuel initiatives in 2012, and SkyNRG, Dutch company originally formed by KLM and Air France in 2009, began buying pre-treated biofuels derived from used cooking oils and refining them into aviation-grade fuel (REN21 2013).

1.6 Closing Remarks

Interest in biofuels began with the oil shocks of the 1970s, but the more rapid development of the biofuel industry in recent years has been primarily driven by mandates, subsidies, climate change concerns, and energy security. Global biofuels production is expected to increase by over 25 % from 2012 to 2018, attaining 2.4 million barrels per day in 2018 and accounting for 3.9 % of global oil demand for road transport (IEA 2013a). As biofuel production continues to expand, investments in capacity expansion and research and development have been made, but the 2008 food crisis emphasized the need to reexamine the consequences of reliance on biofuels. Biofuels remain an important renewable energy resource to substitute for fossil fuels, particularly in the transportation sector, yet it faces short-term production challenges and the sustainability of this option is still far from certain.

As of early 2013, policies promoting the use of renewable fuels in the transport sector have been identified at the national level in 49 countries. Most nations regulate biofuel production by mandates on transport fuel composition, and new national blend mandates. Besides regulations, governments have affected market pricing via national tariffs or subsidies in order to aid developing biofuel industries. However, biofuel support policies in Europe and the United States remain under review, and pressure continues to mount, especially on support for first-generation fuels due to concerns about their potential impacts on food production and on land-use change, biodiversity, and water and potential adverse impacts on climate change mitigation. As a consequence, the annual investment trend on biofuels has been decreasing in recent years.

Appendix: Biofuels Targets and Blending Mandates

Country	Biofuel targets	Blending mandates
Angola		E10
Argentina		E5 and B7
Australia		New South Wales: E4 and B2; Queensland: E5
Belgium		E4 and B4
Brazil		E18-25; B5
Canada		National: E5 and B2. Provincial: E5 and B4 in British Columbia; E5 and B2 in Alberta;E7.5 and B2 in Saskatchewan; E8.5 and B2 in Manitoba; E5 in Ontario
Chile	E5 and B5	
China		E10 in nine provinces
Colombia		E8
Costa Rica		E7 and B20
Dominican Republic	E15 and B2 by 2015	
Ecuador	B2 by 2014 and B17 by 2024	E5 pilot in several provinces
Ethiopia		E5
EU-27	All EU-27 countries are required to meet 10 % of final energy consumption in the transport sector with renewables by 2020	
Fiji		E10 and B5 (voluntary, but mandate expected)
Guatemala		E5
India		E5
Indonesia	10.2 % share in primary energy by 2025	B2.5 and E3
Italy	2,899 ktoe in transport by 2020	
Jamaica		E10
Kenya		E10 in Kisumu
Mexico		E2 in Guadalajara (pilot)
Malawi		E10
Malaysia		B5
Mozambique		E10 in 2012–2015; E15 in 2016–2020; E20 from 2021
Netherlands	5 % in transport fuel mix by 2013; 10 % by 2020	
Nigeria	E10	
Panama		E4 in 2014, E7 in 2015, and E10 in 2016 (mandate expected in 2013)
Paraguay		E24 and B1
Peru		B2 and E7.8
Philippines		E10 and B2
South Africa		E10
South Korea		B2.5
Sudan		E5

(continued)

Country	Biofuel targets	Blending mandates
Spain	Biodiesel 7 % of total energy in transport fuel use by 2012 and 2013; 2,313 ktoe by 2020; Ethanol/bio-ETBE 400 ktoe by 2020; Biofuels 2.7 % of final energy by 2020	
Sri Lanka	20 % supply of all liquid fuels by 2020	
Thailand	Ethanol: 9 million liters/day by 2022; Biodiesel: 5.97 million liters/day by 2022; Advanced biofuels: 25 million liters/day by 2022	E5 and B5
Turkey		E2
United Kingdom	5 % by 2014	
Uganda	720,000 m^3/year produced by 2012; 2.16 million m^3/year produced by 2017	
United States		National: The Renewable Fuels Standard 2 (RFS2) requires 136 billion liters (36 billion gallons) of renewable fuel to be blended annually with transport fuel by 2022. State: E10 in Missouri and Montana; E9–10 in Florida; E10 in Hawaii; E2 and B2 in Louisiana; B4 by 2012, and B5 by 2013 (all by July 1 of the given year) in Massachusetts; E10 and B5, B10 by 2013, and E20 by 2015 in Minnesota; B5 after 1 July 2012 in New Mexico; E10 and B5 in Oregon; B2 1 year after in-state production of biodiesel reaches 40 million gallons, B5 1 year after 100 million gallons, B10 1 year after 200 million gallons, and B20 1 year after 400 million gallons in Pennsylvania; E2 and B2, increasing to B5 180 days after in-state feedstock and oil-seed crushing capacity can meet 3 % requirement in Washington
Uruguay		B5; E5 by 2015
Vietnam	Equivalent to 1 % of domestic petroleum demand by 2015;5 % of demand by 2025	E10
Zambia		E10 and B5
Zimbabwe	10 % share in liquid fuels by 2015	E5, to be raised to E10 and E15

Source: REN21 (2013)

References

Chicon, M. (2011). *U.S. Navy Fuels 'Green Strike Force' with $12 million biofuels contract.* Retrieved 8 December 2011 from http://www.renewableenergyworld.com/rea/news/article/2011/12/u-s-navy-fuels-green-strike-force-with-12-million-biofuels-contract

Chisti, Y. (2008). Biodiesel from algae beats ethanol. *Trends in Biotechnology, 26*(3), 126–31.

Cooper, G. (2012). *2011 Ethanol exports total 1.19 billion gallons; Brazil accounts for one-third of shipments*. Retrieved 10 February 2012 from http://www.ethanolrfa.org/exchange/entry/2011-ethanolexports-total-1.19-billion-gallons-brazil-accounts-for-one-thi/

Doornbosch, R. & Steenblik, R. (2007, September 11–12). Biofuels: is the cure worse than the disease? OECD round table on sustainable development, Paris, France.

European Biodiesel Board (EBB). (2013). *Statistics: The EU biodiesel industry*. Bruxelles, Belgium: Author. Retrieved 19 October 2013 from http://www.ebb-eu.org/stats.php#.

Energy Information Administration (EIA), US Department of Energy. (2009). *Short-term energy outlook supplement: biodiesel supply and consumption in the short-term energy outlook*. Washington, DC: Author.

Energy Information Administration (EIA), US Department of Energy. (2013, January 31). *2012 Brief: U.S. ethanol prices and production lower compared to 2011*. Washington, DC: Author.

Licht, F. O. (2008). World ethanol & biofuels report, vol. 6 No. 10.

Girard, P., & Fallot, A. (2006). Review of existing and emerging technologies for the production of biofuels in developing countries. *Energy for Sustainable Development, 10*(2), 92–108.

Herndon, A. (2012). *Cellulosic biofuel to surge in 2013 as first plants open*. Retrieved 11 December 2012 from Bloomberg.com

IEA. (2006). World energy outlook 2006 (Chapter 14). In *The outlook for biofuels*. Paris, France: International Energy Agency (IEA), OECD.

IEA. (2011). *Technology roadmap—Biofuels for transport*. Paris, France: OECD/IEA.

IEA. (2013a). *Medium-term renewable energy market report 2013*. Paris, France: OECD/IEA.

IEA. (2013b). *Tracking clean energy progress*. Paris, France: OECD/IEA.

IEA/IRENA. (2013). *Australian biofuels investment readiness program. Australian biofuels investment readiness program*. Global renewable energy policies and measures database. Retrieved 18 October 2013 from http://www.iea.org/policiesandmeasures/renewableenergy/

Karekezi, S. et al. (2004). Traditional biomass energy: improving its use and moving to modern energy use, Thematic background paper, Bonn, Renewables 2004, January

Keeney, D., & Nanninga, C. (2008). *Biofuel and global biodiversity*. Minneapolis, MN: Institute for Agriculture and Trade Policy.

Liu, D. (2005, November 13–15). *Chinese development status of bioethanol and biodiesel*. Presentation to biofuels: An energy solution, 2005 world biofuels symposium, Tsinghua University, Beijing.

Nielsen, S. (2012). *Brazil doubling sugar-cane biomass research plan to $988 million*. 19 Retrieved September 2012 from http://www.bloomberg.com/

Organisation for Economic Cooperation and Development/Food and Agriculture Organization of the United Nations (OECD/FAO). (2008). *Agricultural outlook 2008–2017*. Paris, France: Organisation for Economic Cooperation and Development/Food and Agriculture Organization of the United Nations.

Organisation for Economic Cooperation and Development/Food and Agriculture Organization of the United Nations (OECD/FAO). (2012). *Agricultural Outlook 2012–2021*. Paris, France: Author.

Organisation for Economic Cooperation and Development/Food and Agriculture Organization of the United Nations (OECD/FAO). (2013). *Dataset: Agricultural outlook 2013–2022*. Paris, France: Author. Retrieved from http://stats.oecd.org/index.aspx.

RFA. (2008). *Changing the climate: ethanol industry outlook*. Washington, DC: Renewable Fuels Association.

RFA. (2013). *U.S. ethanol capacity*. Washington, DC: Renewable Fuels Association. Retrieved 19 October 2013 from http://www.ethanolrfa.org/pages/statistics#C.

REN21. (2005). *Renewables 2007 global status report* (Paris: REN21 Secretariat and Washington, DC: Worldwatch Institute.

REN21. (2006). *Renewables global status report 2006 update* (Paris: REN21 Secretariat and Washington, DC: Worldwatch Institute).

REN21. (2008). *Renewables 2007 global status report* (Paris: REN21 Secretariat and Washington, DC: Worldwatch Institute).

REN21. (2009). *Renewables global status report: 2009 update* (Paris: REN21 Secretariat).
REN21. (2010). *Renewables global status report: 2010 update* (Paris: REN21 Secretariat).
REN21. (2011). *Renewables global status report: 2011 update* (Paris: REN21 Secretariat).
REN21. (2012). *Renewables global status report: 2012 update* (Paris: REN21 Secretariat).
REN21. (2013). *Renewables 2013 global status report* (Paris: REN21 Secretariat).
Stebbins, C. (2011). *Analysis: U.S. biodiesel on lifesupport, but smiling*. Reuters. Retrieved 3 February 2011 from at http://www.reuters.com/article/2011/02/04/us-usa-bi0diesel-idUSTRE7130H920110204.
USDA. 2011. Argentina Biofuels Annual 2011 U.S. Department of Agriculture Foreign Agricultural Service, Global Agricultural Information Network. (Buenos Aires: 8 July 2011), at http://gain.fas.usda.gov/Recent%20GAIN%20Publications/Biofuels%20 Annual_Buenos%20Aires_Argentina_7-8-2011.pdf.
USAID. 2009. Biofuels in Asia: An Analysis of Sustainability Options. United States Agency for International Development (USAID).
US EPA. (2012, September). *EPA finalizes 2013 biomass-based diesel volume. Regulatory announcement*. Washington, DC: Author.
US EPA. (2013). *Renewable fuel standard*. Washington, DC: Author. Retrieved 18 October 2013 from http://www.epa.gov/otaq/fuels/renewablefuels/.
US DOE. (2008). *World biofuels production: understanding the challenges to meeting the US renewable fuel standard*. Washington, DC: Author.
U.S. Department of Energy (US DOE). (2013). *Alternative fuels data center: Federal laws and incentives*. Washington, DC: Author. Retrieved 18 October 2013 from www.afdc.energy.gov/laws/fed_summary.
World Bank. (2013). *World development indicators*. Retrieved 20 October 2013 from http://data.worldbank.org/indicator/PA.NUS.FCRF
Worldwatch Institute. (2007). *Biofuels for transportation: global potential and implications for sustainable agriculture and energy in the 21st century*. Washington, DC: Worldwatch Institute.

Chapter 2
Biofuel Technologies and Potential

Jay J. Cheng, Govinda R. Timilsina, and David Zilberman

2.1 Introduction

In many cases, the term "biofuels" has been used loosely to include liquid organic fuels that have potential to replace fossil fuels. However, biofuels are classified into different types, such as ethanol, methanol, and biodiesel.

- Ethanol: also known as ethyl alcohol, pure alcohol, grain alcohol, or drinking alcohol, has the basic structure: CH_3CH_2OH. Ethanol is created through the fermentation of sugars or by ethylene hydration. Where food crops (e.g., corn, potato, and sugarcane) all have readily accessible carbohydrates, ethanol has been used by human civilizations for millennia and has served as the first-generation of biofuels.
- Biodiesel: defined by the National Biodiesel Board: the monoalkyl esters of long fatty acids derived from renewable lipid feedstock (vegetable oil or animal fat) for use in standard compression ignition engines. Biodiesel has the basic structure: CH_3CH_2 CH_2 CH_2 … $COOCH_3$, in which "$COOCH_3$" is the ester group. Traditional diesels lack the ester group, and would otherwise appear as: CH_3CH_2 CH_2 CH_2 … CH_3. Esters make the compound slightly more polar and slightly more volatile. Although the first engine powered vehicles ran on vegetable oil, current diesel engines require modifications for a manual switch between an ignition fuel and biodiesel to be used once the engine is running.

J.J. Cheng
North Carolina State University, Department of Biological and Agricultural Engineering, Raleigh, NC 27695, USA

G.R. Timilsina (✉)
The World Bank, 1818 H Street, N.W., Washington, DC 20433, USA
e-mail: gtimilsina@worldbank.org

D. Zilberman
Department of Agricultural and Resource Economics, College of Natural Resources, University of California, Berkeley, Giannini Hall 206, Berkeley, CA 94720-3310, USA
e-mail: zilber11@berkeley.edu

G.R. Timilsina and D. Zilberman (eds.), *The Impacts of Biofuels on the Economy, Environment, and Poverty*, Natural Resource Management and Policy 41,
DOI 10.1007/978-1-4939-0518-8_2,

- Methanol: also known as methyl alcohol or wood alcohol, is the simplest alcohol compound (CH_3OH). As such, methanol has the least carbon to hydrogen ratio than any other liquid fuel. Methanol may be produced from different organic materials, including natural gas, coal, biomass, landfill gas, and crops. “Biomethanol” refers to methanol produced from organic materials, e.g., gasification of organic materials to syngas that is then converted to methanol through conventional synthesis. Through this process, biomass converts at up to 75 % efficiency for fuels.

2.2 Ethanol Production Process

Ethanol is normally produced from feedstocks that have high sugar or starch content like sugarcane (about 16 % soluble sugars), sugar beets (about 18 % sucrose), and corn (about 66 % starch). As opposed to sugar crops, starches require an extra step of saccharification, in which enzymes break down starch into glucose. The sugars are fermented by yeasts or bacteria to convert to ethanol. Through further distillation and dehydration, the ethanol content is then concentrated enough to serve as liquid fuel. When lignocellulosic materials are used as the feedstock, they require pretreatment and hydrolysis to convert the cellulose to glucose before the fermentation. Figure 2.1 illustrates the first- and second-generation ethanol production processes, wherein common feedstock inputs undergo chemical reactions, form intermediate products, and, finally, culminate in usable fuel. Second-generation ethanol production technologies use lignocellulosic biomass as feedstock, including organic waste and residues. These feedstocks are cheap and abundant, although the extra steps and expenses in treatment may not yet be.

2.2.1 Pretreatment

Lignocellulosic materials are composed of mainly cellulose, hemicelluloses, and lignin that form a tight and complex structure. Cellulose is a long-chain homogenous polysaccharide of D-glucose units linked by b-1,4 glycosidic bonds. Hemicellulose is a complex, heterogeneous polymer of 6-carbon and 5-carbon sugars and sugar derivatives. Lignin holds the cellulose and hemicellulose fibers together and provides support to the plants. To release sugars from cellulose and hemicelluloses, the materials have to go through a pretreatment process. Pretreatment generally serves to remove lignin, reduce cellulose crystallinity, and increase porosity of the material so the enzymes can penetrate in the following hydrolysis step. Whether the pretreatment is applied mechanically or chemically or biologically spurs another debate.

- *Physical pretreatment*: involves physical forces to break the lignocellulose structure. Mechanical comminution incorporates chipping, grinding, and milling to reduce the size of lignocellulosic materials down to 0.2–2 mm granules.

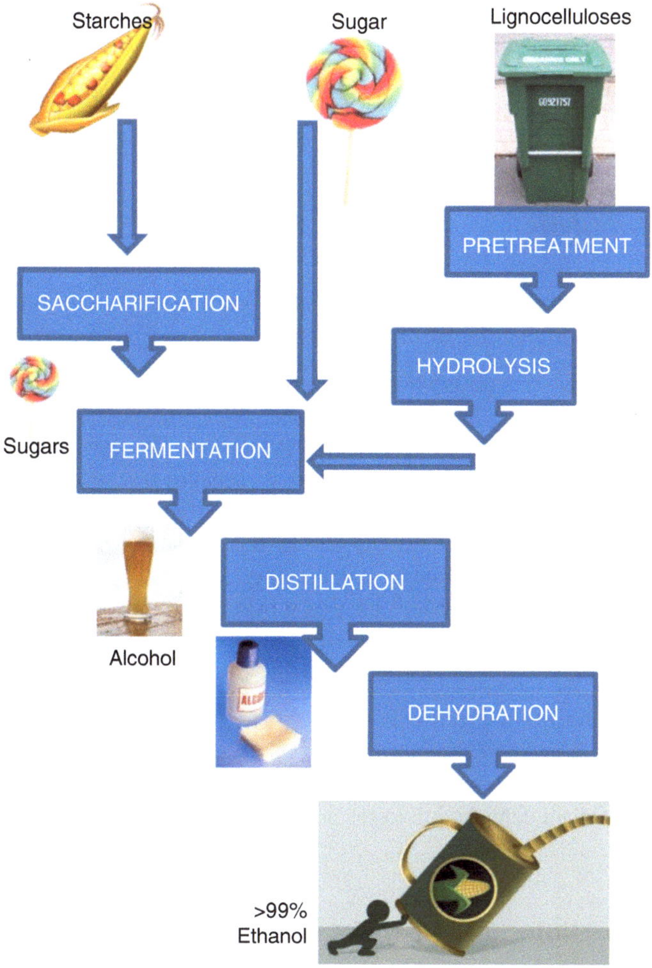

Fig. 2.1 A typical process of ethanol production

Steam explosion applies high temperatures (160–260 °C) and high pressure steam to soak the lignocellulosic material and swiftly release the pressure to atmospheric, causing explosive decompression of the material, and thus separating lignin from carbohydrates and degrading the hemicelluloses. While effective, steam explosion requires large amounts of energy and generates furfural byproducts, which may inhibit the subsequent enzymatic hydrolysis and microbial fermentation.

- *Chemical pretreatment*: subjects lignocelluloses to alkaline or acid hydrolyses at high temperature (100–190 °C). High temperatures promote the chemical reactions with alkaline or acid solution to break the bonds between lignin, cellulose, and hemicellulose. The disadvantage of this treatment is that significant amounts of lignin remain in the mixture and may inhibit cellulase enzymes from binding the substrate, cellulose, during the next stage of process, i.e., enzymatic hydrolysis.

- *Biological pretreatment*: subjects the lignocellulose to microbial fungi (brow-, white-, and soft-rot, which attack hemicellulose or cellulose and lignin). White-rot fungi are the most effective fungi in this process (Fan et al. 1987). This pretreatment is relatively inexpensive, but it usually takes a long time.

2.2.2 *Hydrolysis*

Following pretreatment, the lignocelluloses undergo hydrolysis to release sugars. Again, there are different means to the task:

- *Chemical hydrolysis*: mixing mineral acids with cellulose fibers cause the decomposition of crystalline bundles of cellulose, releasing glucose that can be fermented into ethanol in the following fermentation. The downside to chemical hydrolysis is that some toxic byproducts (e.g., acetate) are created (Clarke 1997), that could be inhibitory to the following fermentation.
- *Biological hydrolysis*: while expensive, the most effective way of releasing soluble sugars from lignocellulose is by using cellulase enzymes—specifically, endoglucanase, exoglucanase, and β-glucosidase. Cellulases are chiefly produced by fungi and bacteria. The specified three enzymes work synergistically to break down the cellulose to the end product, glucose.

Finally, the resulting glucose is ready for fermentation by yeasts. (Common ethanol fermentation yeasts cannot ferment pentose sugars to ethanol.) The rest of the process is the same as the first-generation technology, namely, fermentation, distillation, and then dehydration to reach 99 % fuel-grade ethanol.

2.3 Lignocellulosic Feedstocks

Lignocellulosic feedstocks comprise of three main sources: agricultural residues, forest residues, and herbaceous and woody energy crops. Although the potential yield of lignocellulosic materials is high (67 %), the extraction is quite difficult given the density and entanglement of cellulose and hemicellulose molecules, tightly bound by lignin (Hamelinck et al. 2005).

2.3.1 *Types of Feedstocks*

Agricultural residues: agricultural residues are leftovers in the field after the crops are harvested and usually waste materials that need management. Typical agricultural residues include corn stover, rice straw, wheat straw, etc. The supply of agricultural residues still varies across weather, topography, tillage means, and crop conditions. Exceptions are rice straw and sugarcane bagasse, which could be removed 100 % from agricultural plots. Other residues are estimated to be harvested at only 40 % for biofuel feedstock, given the transportation and opportunity costs for immediate reuse.

Forest residues: another feedstock for cellulosic ethanol is "forest residues" that encompass organic materials left from logging harvests, fuel wood extraction, and mill residues from primary and secondary wood processing (Perlack et al. 2005). While estimating a current extraction potential of 65 gal for fuel per ton of feedstock, National Renewable Energy Laboratory (NREL) had set goals for 94 gal fuel yield per ton of feedstock for 2020. In assessment, however, the conversion rate of forest residues is lower in actuality than in theory. Gaps between theoretical conversion rates are expected to diminish over time. The greatest impediment to realizing theoretical yield is due to inaccessibility. Gathering residues from accessible areas would aggrandize costs to existing wood logging and collection. Furthermore, recovery of forest residues also draws environmental concern for sequestration, soil restoration, and forest regeneration cycles.

Ethanol crops: the biofuel industry relies on a variety of feedstocks: from herbaceous to woody, from arable to residue, from temporal to perennial. Comparative advantages usually dictate what regional inputs are used for production to fuel. However, an overarching comparison might be made for some dedicated feedstocks. Dedicated feedstocks are so chosen for being relatively less input-intensive, easier on soil, and higher yield altogether. Among herbaceous biofuel crops, the following seems quite promising:

- Switchgrass: the Biofuels Feedstock Development Program has identified switchgrass as the most promising bioenergy crop out of 34 herbaceous species (ORNL 2008). Switchgrass is touted based on its relatively low water and input requirements, environmental benefits, and ability to subsist in lower quality lands (Keshwani and Cheng 2009). Switchgrass naturally grows in North America, over a variety of soils as long as water retention is met.
- Miscanthus: native to Asia, is resilient to cold temperatures and low-nitrogen soils. The limitations of this crop involve its narrow genotype (making the crop less adaptable), seasonality, and high costs and long time to establish (requiring wide distribution of rhizome cuttings).
- Bermuda grass: a warm-season perennial grass, naturally grows in Africa, Asia, Australia, and southern Europe. Bermuda grass typically grows in warm climate regions. Its growth season usually starts in the spring at temperatures above 15 °C. The grass reaches its optimum growth at temperatures between 24 and 37 °C. Bermuda grass is among the high growing grasses with an annual yield of 6–8 t (dry matter)/acre/year.

Reed canary grass: is typically used for forage or hay. Although reed canary grass is suited to temperate climates and erosive conditions, once established in wetlands, it can become an invasive species.

Table 2.1 Ethanol yields and field production rates of common lignocellulosic feedstocks

Feedstock		EtOH yield (liter/dry ton)	Production rate (dry ton/acre/year)	EtOH yield (liter/acre/year)
Woody	Poplar tree (hardwood)	360	5–6	1,980
	Pine tree (softwood)	345	3–4	1,208
Herbaceous	Switchgrass	310	6–15	3,255
	Miscanthus	305	6–15	3,203
	Bermuda grass	300	6–10	2,400
Agricultural residues	Corn stover	345	3–5	1,380
	Wheat straw	333	1–3	666
	Rice straw	335	3–4	1,173

Source: Carriquiry et al. (2011)

Woody energy crops generally have higher yield and greater versatility as solid or liquid fuel source (Cheng and Timilsina 2011). The faster growing tree species include:

- Poplar: a genus of 25–35 species of flowering plants. Poplar is attractive for its high-energy output/input ratio, carbon mitigation potential, and growth rate higher than most other woody plants. Potential downsides include wind erosion in stripped fields after harvesting poplar trees for biomass.
- Willow: a genus of about 400 species of deciduous trees and shrubs (Mabberley 1997). Willows are highly cross-fertile and adaptable. While the willow serves many environmental benefits: soil erosion control, windbreak, soil reclamation, and shelterbelt, willow roots are aggressive spatially and tend to drain water resources. In Australia, willows have even become regarded as weeds.
- Pine: a genus of about 115 species of evergreen, coniferous resinous trees. Pines are native to most of the Northern Hemisphere, and have been introduced throughout most temperate and subtropical regions of the world. Pines are among the most commercially important tree species for the production of timber and wood pulp. They are fast-growing softwood trees with biomass yield of about 3–4 t (dry matter)/acre/year.
- Eucalyptus: a genus of more than 700 species concentrated in Australia. Eucalyptus grows very fast with a high biomass yield. Its oil is also multi-purposeful and is highly flammable—which can be a fire hazard in growth—but a highly efficient fuel. Eucalyptus can be an invasive species when transplanted.

The biomass production rate and ethanol yield from various lignocellulosic feedstocks are summarized in Table 2.1.

2.3.2 *Lignocellulosic Ethanol: Research and Development*

Stemming from the complexities of the feedstock and processes for the second-generation fuel ethanol production, advancements must be made in the way of technology and cost structure before these seemingly carbon neutral sources may compete as energy provisions. Primarily, research on lignocellulosic ethanol production has been focused on reducing the costs of the second-generation technologies. Among these weak points are pretreatment technologies, cellulase enzyme efficacy, and co-fermentation of 6- and 5-carbon sugars (e.g., glucose, xylose) to boost output.

Feedstock development: genetic modifications for corn stover, switchgrass, and poplar trees have been pursued. Lower lignin would reduce pretreatment costs. Higher cellulose would aggrandize the yield. So far, there has been some success in reducing the lignin content in aspen trees, but there remains work to be done.

Pretreatment technology: pretreatment technologies use most energy for having to hold high temperatures or high pressure. In order to create high temperatures, the machine capital and utility usage is significantly higher than technologies for first-generation ethanol production (Sassner et al. 2008). Thus, research and development is pursuing lower temperature, lower pressure models for cost reduction.

Enzyme optimization: while cellulase enzymes have lowered in cost over the last decade, these are not yet competitive with the enzymes (e.g., amylases) used in the first-generation starch-to-ethanol platform (Carriquiry et al. 2011). Higher efficiency cellulases are under exploration to make the enzymatic hydrolysis stage of production more cost-effective. Also, new technologies are needed to lower the cost of cellulose sources.

Co-fermentation: since the main products of hydrolysis are glucose (from cellulose) and xylose (from hemicelluloses), there are two sugars that could yield energy. Current technologies have mastered the conversion of glucose by yeasts and bacteria, yet the conversion of xylose to ethanol remains problematic. Research efforts are needed to engineer microorganisms to efficiently convert glucose and xylose to ethanol, synergistically, to maximize ethanol output.

2.3.3 Pilot Facilities for Lignocellulosic Ethanol Production

In spite of existing barriers for the commercial deployment of advanced bioethanol, some companies have begun pilot-scale lignocellulosic ethanol production in Europe, North America, and Asia. Table 2.2 summarizes these factories' ownerships, location, feedstock, and output capacity.

Established in 2004, SEKAB has been continuously operated to generate 300–400 L of ethanol per day. With forest residue from pine trees, the factory uses acid pretreatment, enzymatic hydrolysis, and yeast fermentation. The European Union funds this pilot project with the hopes of extending production with bagasse from sugarcane, wheat straw, corn stover, energy grass, and organic waste materials.

Established in 2002, Arkenol generates 100–300 L of ethanol per day. Inputting mixed forest residues, the facility uses a distinct method of sulfuric acid to convert carbohydrates to simple sugars. The sugar precipitates are filtered out from the liquid solution. Then, a recombinant bacterium (*Z. mobilis*, developed by the National Renewable Energy Laboratory of the US Department of Energy) is applied along with *S. cereviscae* for fermentation. Finally, the ethanol product is purified through

Table 2.2 Examples of pilot-scale lignocellulosic ethanol production facilities

Company	Location	Raw materials	Capacity
SEKAB E-Technology	Örnsköldsvik, Sweden	Wood chips from pine trees	300–400 L of ethanol per day
Arkenol	Izumi, Japan, and Irvine, CA, USA	Mixed waste wood chips of cedar, pine, and hemlock	100–300 L of ethanol per day
Iogen	Ottawa, ON, Canada	Wheat straw	Process 40 t of wheat straw to ethanol per day

Source: Carriquiry et al. (2011)

distillation (via membrane) and dehydration that cost less to operate than a conventional system. Arkenol has achieved a conversion efficiency rate of 70 % for cellulose, and may be optimized to 80 %. Sulfuric acid is able to be recovered and re-applied to treatment at 97 %. The facility is also able to channel the heat from lignin combustion to provide power during the distillation stage.

Established in 2004, Iogen was designed to handle 40 t of wheat straw per day and to manufacture enzymes in an adjacent building. The company has produced cellulosic ethanol for international demonstrations and for the national flexible fuel market, per request of the Canadian government. Currently, Iogen is collaborating with the US and Canadian Energy Agencies to develop a larger scale lignocellulosic processing facility.

More industrial bodies have also gotten involved in lignocellulosic ethanol pilots, including: BlueFire Ethanol and Colusa Biomass Energy Cooperation (CBEC) in California, USA, Brelsford Engineering in Montana, USA, and Masada resource Group, LLC in Alabama, USA. Other industrial companies pursue developing efficacious cellulase enzymes and hydrolysis technologies.

2.4 Biodiesel Production Process

In traditional (first-generation) biodiesel production, vegetable oils (known as biolipids) and alcohol (usually methanol) are subjected to a process called transesterfication (swapping an ester group for an alcohol group) which is catalyzed by acid, alkaline, or lipase enzymes. The effect is that oil or fats are converted to crude biodiesel with glycerin as the byproduct. The product and byproducts continue to be refined until the diesel is concentrated for commercial use. A typical biodiesel production diagram is shown in Fig. 2.2.

2.4.1 Types of Biodiesel Feedstocks

The main feedstocks used to produce biodiesel are discussed below:

Biodiesel crops: among the production efforts in the world, the main biodiesel feedstocks incorporate oil from soybean, rapeseed, canola, sunflower, corn, palm

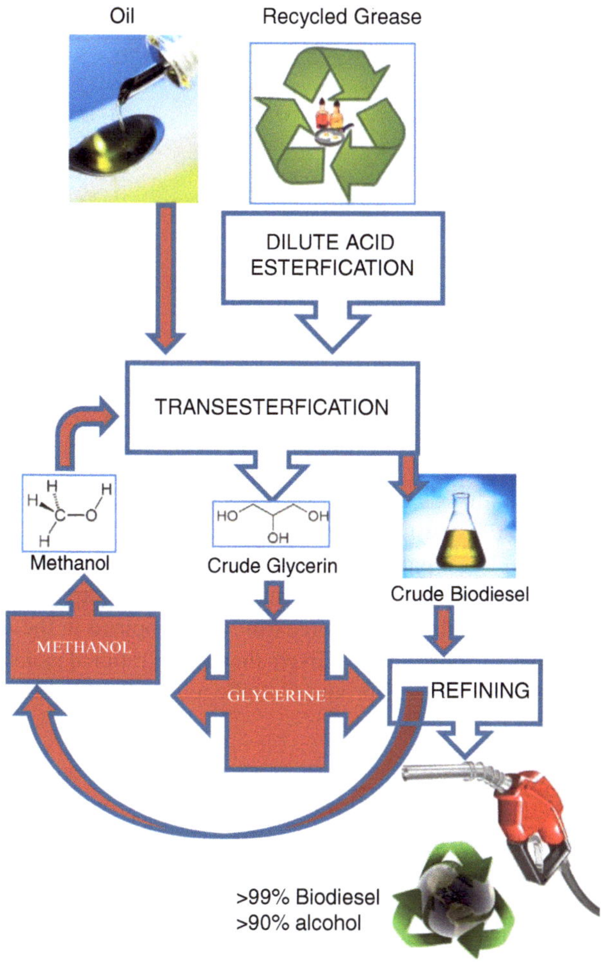

Fig. 2.2 A typical biodiesel production process

kernels, animal fats, and recycled greases. In India and Africa, *Jatropha* has also been found to be an effective oil source as well:

- *Jatropha*: a genus of about 175 succulent plants, shrubs, and trees. The species *Jatropha curcas* had been named by Goldman Sachs as one of the best candidates (Currie 2007) for biodiesel production. Jatropha has an average yield of 34.4 % of seeds that have 27–40 % oil (Achten et al. 2007). The plant is resilient against drought and pests, although *Jatropha* production has not yet been adapted for cultivation.
- Soybean: the seed of this perennial plant has high nutritional value with 40 % proteins, 34 % carbohydrates, 21 % oil, 5 % ash, and trace amounts of phospholipids, sterols, and minerals. As such, soybean serves as feedstock for human consumption, livestock consumption, and fuel.

Table 2.3 Some high oil microalgae

Microalgae	Oil content (as % dry weight)
Botryococcus braunii	25–75
Cylindrotheca sp.	16–37
Isochrysis sp.	25–33
Nannochloris sp.	20–35
Nannochloropsis sp.	31–68
Neochloris oleoabundans	35–54
Nitzchia sp.	45–47
Schizochytrium sp.	50–77

Source: Chisti (2007)

- Rapeseed: *Brassica napus* is widely cultivated in Europe for forage and for vegetable oil. The seed of the plant has a very high content of oil (approximately 40 %). Engineered varieties of rapeseed from Canada are called canola.

Vegetable oils' viability as a fossil fuel replacement is challenged by land constraint. It has been estimated that 24 % of the existing cropland in the US would need to be reassigned to oil palm (as a high yield oil crop) in order to replace just 50 % of the transportation sector's fuel (Chisti 2007). In contrast, replacing fossil fuels with soybean oil would require three times the cropland in the US to be reallocated toward soybean cultivation. Given that these rational constraints would make feedstock undermine food production altogether, the biodiesel may not be sustainable based on food crops alone.

Waste oil: second-generation biodiesel production incorporates waste oil—for example, leftover or recycled oils from the food industry, restaurants, and households. Waste oils contain significantly more free fatty acids, more water, and less triglycerides than fresh vegetable oils. As such, waste oils require pretreatment before transesterfication may begin. In 2007, waste oil was measured at 15 million tons from the US (which generated 10 million tons), EU, China, Canada, Japan, and Malaysia (Gui et al. 2008).

Microalgae: certain types of microalgae have fast growth rates (doubling mass within 24 h under normal growth conditions) and high oil content (up to 75 % of its dry weight). Table 2.3 shows some high-oil microalgae and their oil content. These advantages make microalgae a potentially promising feedstock for biodiesel production. Some microalgae can produce oil in a very effective way with annual oil production ranging up to 58,700– 136,900 L per hectare.

Commercial production of microalgae is conducted in either open ponds or in closed photobioreactors under autotrophic or heterotrophic conditions. Figure 2.3 shows a diagram of microalgae cultivation for biodiesel production. Autotroph means self-feeding and refers to microalgae that have the ability to take light as their energy source and inorganic CO_2 as the carbon source to produce organic compounds (e.g., carbohydrates, fats, proteins) by photosynthesis (Odum and Barrett 2004).

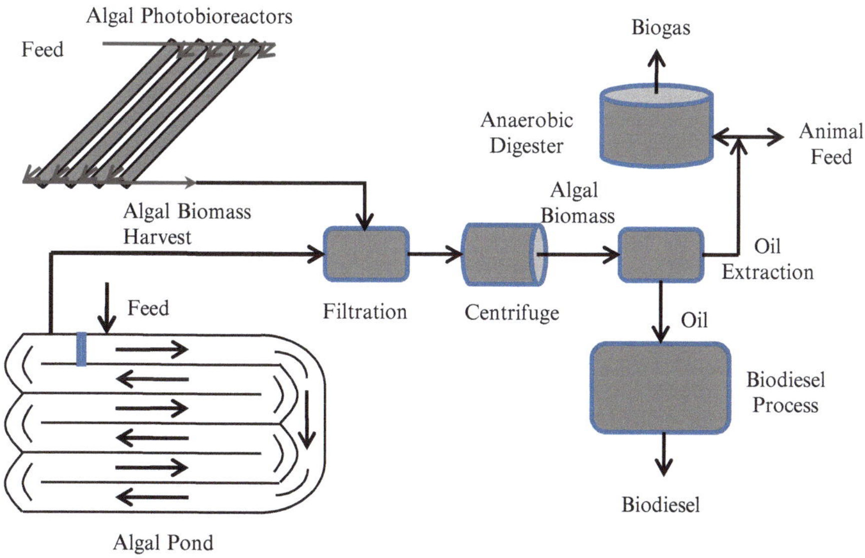

Fig. 2.3 Microalgae cultivation for biodiesel production. *Source*: Cheng and Timilsina (2011)

Heterotrophs may use a combination of photosynthesis or chemosynthesis with organic compounds as their carbon and energy sources. Typically, autotrophic microalgae are used for biodiesel production, given the zero net carbon dioxide emission. Yet, some heterotrophic microalgae produce higher oil content: when *Chlorella protothecoides* is grown autotrophically, it yielded 14.5 % oil (dry weight); grown heterotropically, it produced 55.2 % oil on dry weight basis (Miao and Wu 2006).

Open pond growth systems are usually rather shallow (0.3–0.5 m or 12–18 in.). Microalgae density is usually lower, and the potential for contamination by wild algae is higher in open ponds than in bioreactors. The water in ponds must be circulated to prevent algae from settling. However, ponds are cheaper (only 10–14 % of the cost of photobioreactors) given natural energy sources, nutrients, de-oxygenation, and temperature adjustments.

Photobioreactors involve a tube structure made of glass or plastic, which allows light penetration. Nutrients and water are cycled through to keep the microalgae from settling against the surfaces. These photobioreactor settings provide a pure culture for microalgae and are used for high-oil specie types. Temperatures are maintained at 20–30 °C through heat exchange in the tube system. Oxygen generated is also released to prevent the microalgae from suffocating itself. Photobioreactors are able to produce monocultures in its purified setting and to raise oil content by growing for longer period of time. Table 2.4 demonstrates a comparison of open ponds with photobioreactors for microalgae production.

Table 2.4 Open pond vs. photobioreactor cultivations of microalgae

Open ponds	Photobioreactors
Low volumetric productivity	High volumetric productivity
Low algal biomass concentration	High algal biomass concentration
Easy contamination from wild algae, bacteria	Pure algal culture can be maintained
Higher area requirement	Low area requirement
Low equipment and operational costs	High equipment and operational costs

Source: Cheng and Timilsina (2011)

2.4.2 *Biodiesel: Research and Development*

Selective breeding: oil-rich microalgae species may be selected and engineered for optimal yields. Meanwhile, other biological advantages—such as rapid growth—should also be preserved. Oil-rich species of microalgae currently are outcompeted by wild algae and bacteria. If the desirable oil-rich microalgae may be better engineered towards survival, there would be prevented losses in open pond cultures. Alternatively, some means of inhibiting wild algal or bacterial activities may be considered for adapting ponds. Since maintenance costs of photobioreactors are very high, one way to reduce the costs is to develop microalgae with greater tolerance to temperatures (expanding its range outside 20–30 °C) or to oxygen levels. Higher tolerances would make temperature regulation or fluid circulation requirements less frequent.

Growth systems: both photobioreactors and open ponds have weaknesses and room for improvement. Photoreactors have greater surface areas that algae can potentially cling to. If microalgae settle upon these surfaces, then light will fail to penetrate the glass or plastic walls and there is decreased yield of microalgal biomass. Research is being carried on for materials and surfaces that prevent algae adhesion. Open ponds' main advantage is little addition of inputs required, but the natural environment threatens the purity and high yield of microalgal biomass. Some researchers have suggested greenhouse pond as a means of preventing contamination while still relying on natural sunlight and fluid circulation.

Harvesting: current microalgae harvesting technologies induce coagulation, filtration, and centrifugation at expensive rates. More innovative and effective harvesting systems need to be explored for this rather large cost component.

Biorefinery: perhaps applying biorefinery strategies could improve the economics of microalgal biodiesel production. For example, microalgal biomass has oils that can be extracted for biodiesel production, the residual carbohydrates, proteins, and minor nutrients may be utilized for value-added byproducts. The byproducts are highly valuable as nutraceutical products, animal feed, biogas production (via anaerobic digestion), or organic chemicals. Efficient allocation of byproducts could not only mitigate costs (from additional revenues) but also eliminate material wastes.

2.4.3 *Microalgal Biodiesel: Pilot Facilities*

Several pilot-scale microalgal biodiesel production facilities have been in operation, include:

- Aurora Biofuels, Inc.: a bioenergy firm in California, USA has established a production facility in Florida, USA since 2007. Oil-rich microalgae are grown on seawater ponds on non-arable land. Efficient technologies are used for harvesting and oil extraction processes.
- Old Dominion University: science and engineering personnel have constructed an algal farm in Virginia, USA in 2008. Treated wastewater runs into a one-acre open pond to produce algal biomass. About 3,000 gal of biodiesel fuel have been made from extracted algal oils.
- Renewable Energy Group (REG): a biodiesel production company in Iowa, USA had developed technology to produce high quality microalgal biodiesel since 2008. The pretreatment technology could refine crude oils from a variety of microalgae. The purified oil then undergoes transesterification to produce biodiesel.

2.5 Production of Biomethanol and Fischer–Tropsch Fuels

2.5.1 *Gasification*

Whereas in producing bioethanol and biodiesel, only certain portions of plant biomass are used as feedstock, the gasification process can convert all solid waste into gaseous fuel. Gasification is achieved by reacting carbonaceous material at controlled air flow and over 700 °C. The output gases (called synthesis gases or syngas or producer gases) are renewable fuels that have the energy to drive turbines to produce electricity. The syngases can also be used to synthesize methanol and Fischer–Tropsch fuels. A typical gasification process is shown in Fig. 2.4.

Gasification was historically a means of oxidizing coal in the early 1800s. China continues to gasify its abundant coal for gaseous fuels, while other nations gasify organic residues and biomass into syngas. A typical gasification process takes four steps:

1. *Heating and drying*: occurs as high temperatures permeate the solid biomass, causing moisture to escape from the biomaterial. Water and other fluids evaporate at high temperature, but no chemical reactions occur at this stage.
2. *Pyrolysis*: in the absence of oxygen, the organic biomass decomposes. Hemicellulose decomposes first (when 225–325 °C is reached), followed by cellulose (at 300–400 °C), and then other biomass components at higher temperatures. Lignin is the most difficult component to decompose (at up to 500 °C). The main products of pyrolysis are volatile gases (H_2, CO, CO_2, CH_4, and other light hydrocarbons) and char (porous carbonaceous residue).

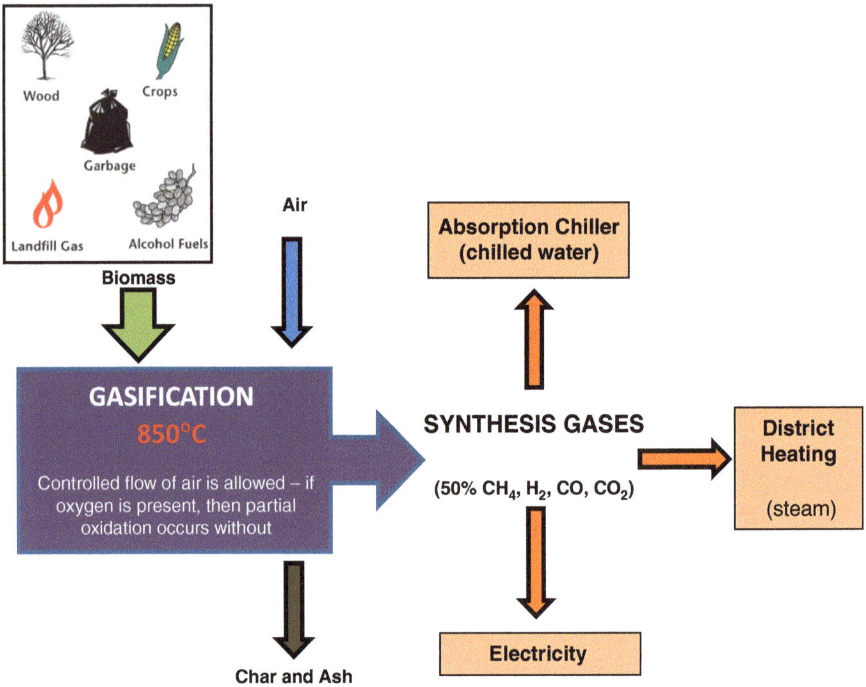

Fig. 2.4 A typical gasification process. *Source*: University of Minnesota (2008)

3. *Gas–solid reactions*: the volatile gases and char can react with each other. The major interactions are presented stoichiometrically, with the change in energy in the following reactions:

$$C + ½O_2 \leftrightarrow CO \qquad \Delta H = -110\,kJ \tag{2.1}$$

$$C + O_2 \leftrightarrow CO_2 \qquad \Delta H = -394\,kJ \tag{2.2}$$

$$C + 2H_2 \leftrightarrow CH_4 \qquad \Delta H = -75\,kJ \tag{2.3}$$

$$C + CO_2 \leftrightarrow 2CO \qquad \Delta H = 172\,kJ \tag{2.4}$$

$$C + H_2O \leftrightarrow H_2 + CO \qquad \Delta H = 131\,kJ \tag{2.5}$$

4. *Gas phase reaction*: as the number of volatile gases increase in the chamber, there are gas–gas reactions. Major interactions are as follows:

$$CO + H_2O \leftrightarrow H_2 + CO_2 \qquad \Delta H = -41\,kJ \tag{2.6}$$

$$CO + 3H_2 \leftrightarrow CH_4 + H_2O \qquad \Delta H = -206\,kJ \tag{2.7a}$$

The final contents of the gasification process depend on the amount of oxygen and steam that are introduced into the system. Temperature and duration of time in the chamber also affect the products of the gasification. From the major reactions listed above, the products include: CO, CO_2, H_2, H_2O, CH_4, and N_2. Also produced, in lesser quantities, are some byproducts (NH_3, H_2S, HCl) and tars that are altogether considered contaminants in the chamber. To purify the syngas products, water is sprayed into the mix, so that the contaminants precipitate out and can be removed.

2.5.2 Biomethanol Production from Syngas

Biomethanol production can occur from the final gas–gas elements present in the gasification chamber. At temperature and pressure of 260 °C and 100 psi, respectively, steam and copper-zinc oxide will catalyze the reaction of CO and H_2. The reaction shown below is the alternate version of reaction (2.7a) under these conditions:

$$CO_2 + 3H_2 \leftrightarrow CH_3OH + H_2O \tag{2.7b}$$

Biomethanol, produced from syngases, can be used directly as liquid fuel, or applied as the alcohol in biodiesel production, or may be added into processed biodiesels (such as dimethyl ether). Dimethyl ether is created from two methanol molecules shown as follows:

$$2CH_3OH \leftrightarrow CH_3OCH_3 + H_2O \tag{2.8}$$

2.5.3 Fischer–Tropsch Fuels

In 1920s, Franz Fischer and Hans Tropsch invented and patented a process to convert coal into liquid fuels. The Fischer–Tropsch (F–T) process is a number of chemical reactions that convert CO and H_2 into liquid alkane compounds. The resultant liquid fuel has low amounts of sulfur, which burns cleaner than regular diesel. Catalyzed by transitional metals (cobalt, iron, ruthenium, and nickel), the F–T process occurs from syngas components at a ratio presented in reaction (2.9), to produce a chain of alkanes and water:

$$nCO + (2n+1)H_2 \leftrightarrow C_nH_{(2n+2)} + nH_2O \tag{2.9}$$

The F–T products depend in large part on the temperature, pressure, and catalysts present. For example, relatively lower temperatures favor higher alkanes—such as gasoline, diesel, and jet fuel. Relatively higher temperatures favor methane formation. Cobalt and iron catalysts are used in the F–T process to expedite long-chains of alkanes. Nickel, when present as a catalyst, tends to help methane formation.

2.5.4 *Biomethanol and F–T Fuels: Research and Development*

Gasification (or the syngas product) is central to the biomethanol production and F–T processes. As such, the biomethanol and F–T fuels share the high costs of capital, operation, and maintenance of gasifiers.

Feedstock: traditional feedstocks of coal and natural gas have been well-documented for gasification. More studies need to be done for the gasification of biomass feedstocks and the long-term impacts of the syngases. On another note, biomass has a lower energy density than coal, resulting in a lower yield of syngases. An optimal feedstock that maximizes the yield to toxin ratio might be a topic of research.

Facility maintenance: the impurities and byproducts of gasification (NH_3, H_2S, HCl, tars, and higher hydrocarbons) require expedient, cost-effective removal. Many of the catalysts for methanol and F–T reactions (e.g., cobalt, iron) are highly sensitive to the impurities. The balance of ongoing operations and system maintenance requires further insight and development.

Fuel transportation: there are usually long distance costs incurred to transport the biomass to an F–T fuel plant and, then, to deliver the product fuels to commercial locations. Most efficient and risk-averse means would be a natural gas pipeline. Therefore, civic planning and economic analyses should be made as to gasifier or plant locations within pipeline distance of feedstock resources.

2.6 Closing Remarks

Lignocellulosic ethanol, microalgal biodiesel, and Fischer–Tropsch fuels each have the potential to contribute to transportation sector's consumption in a more environmentally friendly way. Lignocellulosic materials are abundant and already existing in most regions of the world. These feedstocks do not necessarily compete for arable land, thus, not competing with food and animal feed. Microalgae have demonstrated the ability to produce oil with many folds higher yield than most oil plants. Microalgae also have the advantage of requiring little land area and not necessarily arable land. Commercial production of microalgae would also benefit large-scale carbon dioxide (a greenhouse gas) absorption from the air and replenishment of oxygen. F–T fuels are able to re-use almost any organic waste for the generation of energy.

Current setbacks in advanced biofuel technologies involve several technical issues. Lignocellulosic ethanol, for example, has trouble converting xylose and carbon sugar byproducts into ethanol. Research efforts must address these weaknesses in order to improve the competitiveness of second-generation ethanol. In the case of microalgal biodiesel, the main challenges arise from biomass production costs (mainly in purification and harvest). Microalgal byproducts also require investigation for fuel conversion. The F–T fuel technology suffers a negative net energy yield due to the high energy usage in processing. More efficient gasification might be uncovered through study of low-cost, high-efficiency catalysts, feedstock/fuel transportation, removal of contaminants, and syngas feedback loop to power the gasifier system.

References

Achten, W. M. J., Mathijs, E., Verchot, L., Singh, V. P., Aerts, R., & Muys, B. (2007). Jatropha biodiesel fueling sustainability? *Biofuels, Bioproducts and Biorefining, 1*(4), 283–291.

Carriquiry, M. A., Du, X., & Timilsina, G. R. (2011). Second-generation biofuels: Economics and policies. *Energy Policy, 39*(7), 4222–4234.

Cheng, J. J., & Timilsina, G. R. (2011). Status and barriers of advanced biofuel technologies: A review. *Renewable Energy, 36*, 3541–3549.

Chisti, Y. (2007). Biodiesel from microalgae. *Biotechnology Advances, 25*, 294–306.

Clarke, A. J. (1997). *Biodegradation of cellulose* (pp. 1–272). Lancaster, PA: Technomic.

Currie, J. (2007). Food, feed and fuels: an outlook on the agriculture, livestock and biofuel markets. The Goldman Sachs Group. Retrieved March 2007 from http://www.gceholdings.com/pdf/goldmanreportfoodfeedfuel.pdf

Fan, L. T., Gharpuray, M. M., & Lee, Y.-H. (1987). *Cellulose hydrolysis biotechnology monographs* (p. 57). Berlin, Germany: Springer.

Gui, M. M., Lee, K. T., & Bhatia, S. (2008). Feasibility of edible oil vs. non-edible oil vs. waste edible oil as biodiesel feedstock. *Energy, 33*(11), 1646.

Hamelinck, C., Hooijdonk, G., & Faaij, A. (2005). Ethanol from lignocellulosic biomass: Techno-economic performance in short-, middle- and long-term. *Biomass and Bioenergy, 28*, 384–410.

Keshwani, D., & Cheng, J. J. (2009). Switchgrass for bioethanol and other value-added applications: A review. *Bioresource Technology, 100*, 1515–1523.

Mabberley, D. J. (1997). *The plant book* (2nd ed.). Cambridge, England: Cambridge University Press.

Miao, X., & Wu, Q. (2006). Biodiesel production from heterotrophic microalgal oil. *Bioresource Technology, 97*, 841–846.

Oak Ridge National Laboratory. (ORNL). (2008). Exploring potential US switchgrass production for lignocellulosic ethanol. ORDL/TM-2007/183.

Odum, E. P., & Barrett, G. W. (2004). *Fundamentals of ecology* (5th ed., p. 598). Belmont, CA: Brooks Cole.

Perlack, R., Wright, L., Turhollow, A., Graham, R., Stokes, B., & Erbach, D. (2005, April). *Biomass as feedstock for a bioenergy and bioproducts industry: The technical feasibility of a billion-ton annual supply*. Department of Energy/GO-102005-2135.

Sassner, P., Galbe, M., & Zacchi, G. (2008). Techno-economic evaluation of bioethanol production from three different lignocellulosic materials. *Biomass and Bioenergy, 32*, 422–430.

Chapter 3
Production Costs of Biofuels

Miguel A. Carriquiry, Xiaodong Du, and Govinda R. Timilsina

3.1 Introduction

The production cost of biofuels is one of the key determinants of the commercial viability of biofuels and its social costs of promotion through fiscal stimuli and regulations. Estimates of production costs for different types of biofuels vary widely and are evolving over time. The sources of variability depend on the category/feedstock/production technology. The costs of first generation biofuels, whose production technologies are matured with commercial production, are influenced mostly by costs of feedstock. In the case of corn-based ethanol, for example, feedstock accounts for about 70 % of the total production costs (see Fig. 3.1). For biodiesel, the share of feedstock in total costs of production is even higher, reaching 85–90 %. The recent price volatility in agricultural commodities further contributed to the higher costs of biofuels. In the case of second generation biofuels, much less is known in terms of both process technologies and costs, as there is little experience on commercial production. The available costs are ex-ante estimates with assumptions changing in each estimate. Also, technology pathways for converting cellulosic biomass into biofuels are associated with technical and cost uncertainties.

M.A. Carriquiry
Universidad de la Republica, Department of Economics,
Joaquin Requena 1375, Montevideo 11200, Uruguay
e-mail: carriquiry@iecon.ccee.edu.uy

X. Du
Department of Agricultural and Applied Economics, University of Wisconsin-Madison,
427 Lorch Street, Madison, WI 53706, USA
e-mail: xdu23@wisc.edu

G.R. Timilsina (✉)
The World Bank, 1818 H Street, N.W, Washington, DC 20433, USA
e-mail: gtimilsina@worldbank.org

G.R. Timilsina and D. Zilberman (eds.), *The Impacts of Biofuels on the Economy, Environment, and Poverty*, Natural Resource Management and Policy 41,
DOI 10.1007/978-1-4939-0518-8_3, © Springer Science+Business Media New York 2014

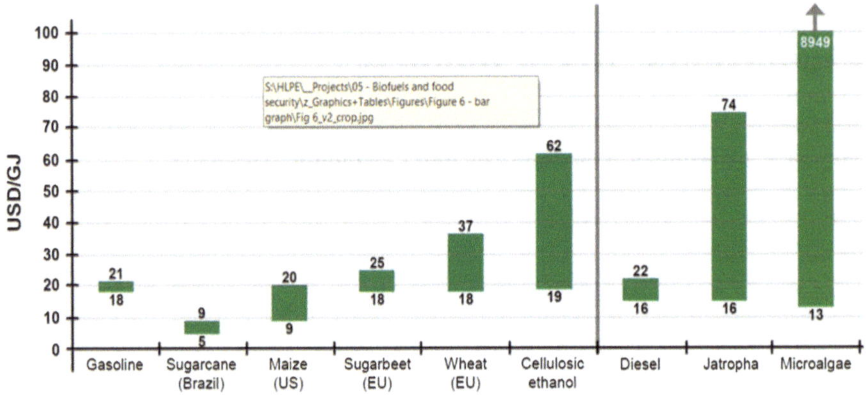

Fig. 3.1 Biofuel production costs from various feedstocks. *Source*: HLPE (2013) and Carriquiry et al. (2011). *Note*: co-product credits are not included for maize and wheat ethanol in the study. *GJ* giga joules

In this chapter, we review the literature on production costs of biofuels. We start with an overview of key issues related to the cost estimation of first generation biofuels (i.e., ethanol and biodiesel). We then move to the review of production costs of second generation biofuels, highlighting the need for new and grounded data on production processes and costs of plants that operate at commercial scale. To date, most cost estimates of second generation biofuels are ex-ante engineering estimates. Data from commercial scale operations is necessary to provide more realistic cost insights. Carriquiry et al. (2011) provided more detailed accounts of costs and technical potential of second generation biofuels at the global level.

3.2 First Generation Biofuels

As indicated in the introduction, production costs of first generation biofuels are critically dependent on the price of agricultural products, in particular cereals and vegetable oils. The costs of biodiesel are more sensitive to the levels of feedstock prices (vegetable oils or animal fats) than the costs of ethanol.

The relative competitiveness of first generation biofuels against the fossil fuels they displace also depends on the point in time at which the comparison is made because prices of fossil fuels and biofuel feedstock change every day. Brazilian sugarcane, ethanol, and US corn ethanol appeared to be cost competitive with gasoline in the market for a wide range of the crude oil prices observed since the mid to late 2000s and value of bi-products such as heat and electricity produced from bagasse in Brazil and Dried Distillers Grain Solubles (DDGS) in the United States (HLPE 2013). In general, productions of biofuels are mainly triggered through government interventions (subsidies or consumption mandates). Costs of biodiesel, in general, are higher than that of diesel.

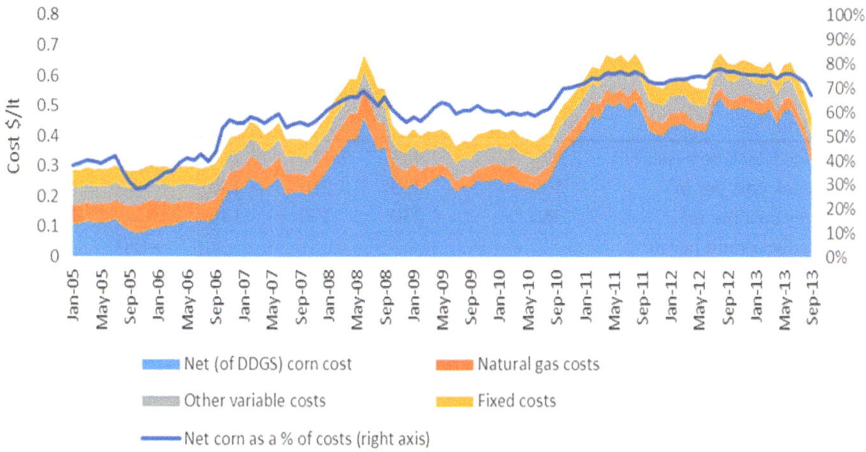

Fig. 3.2 Monthly costs of corn ethanol production of representative plants in Iowa, US. *Source*: http://www.extension.iastate.edu/agdm/reoutlook.html

3.2.1 Ethanol

In the past, Brazil was widely seen as the low-cost producer of ethanol. In fact, by the mid-2000 some authors observed that Brazil's sugarcane-based ethanol was the only biofuel that could compete (without subsidies) with oil prices below US$70 per barrel (Doornbosch and Steenblik 2007; Worldwatch Institute 2007). Brazil's ethanol production costs, at US$0.18/L were lower than those of cereal or sugarcane-based ethanol produced elsewhere. Chinese ethanol costs, depending on the feedstock (sweet sorghum or cassava) were listed in the US$0.28–0.46/L range, similarly molasses-based ethanol costs were reported to be US$0.44/L. The corn ethanol industry in the United States has grown with intensive policy support from federal and state governments. After a significant wave of capital investment and expansion through around 2009, by January 2013, a total of 211 ethanol plants in 28 states have the annual production capacity of 14.7 billion gallons. Over the period of 1983–2005, feedstock cost for corn ethanol decreased by 70.5 % from $631/m^3$ to $186/m^3$ and industrial processing cost declined by 51.8 % from $272/m^3$ to $131/m^3$. These two together brings down the total production cost of corn ethanol by about 65 % from $903/m^3$ to $317/m^3$ (Hettinga et al. 2009). Feedstock cost reduction was largely induced by corn yield improvement and increase in average farm size, reduced enzyme costs, better fermentation technologies, distillation and dehydration, heat integration, and automation (Hettinga et al. 2009). Learning-by-doing and market competition from imported ethanol are also found to play an important role in the reduction of processing cost (Chen and Khanna 2012). The cost trend has been, however, reversed since 2005 due to increased feedstock prices. Figure 3.2 shows monthly costs of production of corn ethanol, for a representative plant in the United States, as tracked by Iowa State University for the January 2005 to September

Table 3.1 Cost comparison between sugarcane ethanol in Brazil and corn ethanol in US

Cost item	Cost of sugarcane Brazilian R$	US$, when 1 US$ = R$1.55	R$2.15	R$2.62	Cost of corn (US$)
Total feedstock cost (per ha)	3,074	1,983	1,430	1,173	1,443
Feedstock cost per m^3	500	320	230	190	340
Refinery cost per m^3	360	230	170	140	190
Co-product credit per m^3					−120
Total production cost per m^3	860	550	400	330	410

Source: Based on Crago et al. (2010)

2013 period. The chart shows that ethanol cost increased significantly since 2005. Increasing and highly volatile feedstock costs have contributed to the increasing trends of ethanol prices. The share of feedstock costs (net of co-product credits) have increased from 40 to 80 % during the 2005–2013 period.

Like in the cost of corn ethanol in the United States, the production costs of sugarcane ethanol in Brazil also dropped until a few years back due to economies of scale and increased yield (van den Wall Bake et al. 2009; Crago et al. 2010). The share of feedstock costs in the total production costs in Brazil was reported to be much lower than US corn ethanol (about 40 % in 2005 by van den Wall Bake et al. 2009 and approximately 30 % for 2006–2008 by Crago et al. 2010). The itemized cost comparison between sugarcane and corn ethanol is provided in Table 3.1. The other factor behind lower ethanol costs was the fuel savings from electricity co-generation in sugar mills. As can be seen from the table, the cost advantages of ethanol production in Brazil and the United States depend, among other things, on the exchange rate at the time the comparison is made. Crago et al. (2010) also calculate the costs of corn ethanol and Brazilian sugarcane ethanol in the United States. For this purpose, they add transportation costs to the Brazilian ethanol. Results of this comparison (for an exchange rate of R$2.15 per US$) are presented in Fig. 3.3. When transportation costs are accounted for, US corn ethanol would be cheaper than Brazilian ethanol in the United States.

3.2.2 *Biodiesel*

Like in the case of ethanol, the estimations of biodiesel costs are sensitive to time of estimation and type of feedstock. In some early studies (e.g., as referred in Worldwatch Institute, 2007) production costs of biodiesel from rapeseed oil ranged US$0.7–1.0/L. The lowest cost of biodiesel production was reported in China, at US$0.21/L from cooking oil. In general, used oils are the cheapest to procure (though might have high collection costs and/or require additional processing to produce biodiesel), followed by palm oil, soybean oil, and rapeseed oil. Table 3.2 compares costs of biodiesel in different geographical locations as reported by Yusuf et al. (2011). Similarly, a comparison made by Carriquiry et al. (2011) of production costs of various biodiesel feedstocks are/is presented in Fig. 3.4.

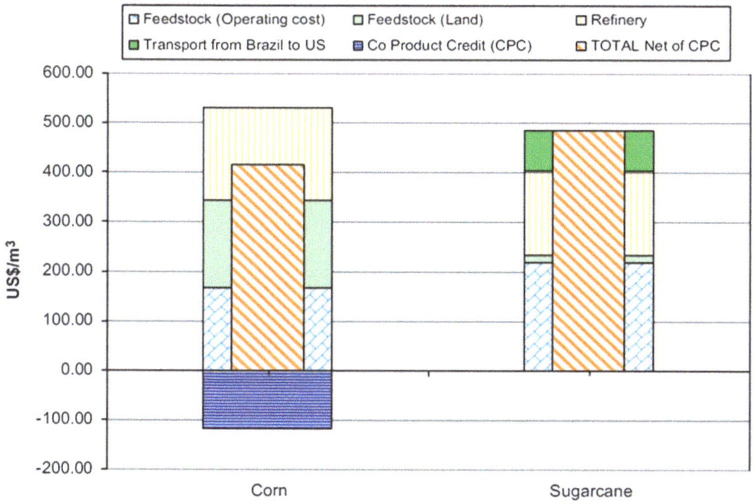

Fig. 3.3 Costs of ethanol production in Brazil and the US. *Source*: Crago et al. (2010)

Table 3.2 Production cost ($/t) of biodiesel from various feedstock in different geographical locations

Item	Palm oil from Malaysia	Rapeseed oil from EU	Soybean oil from the US
Feedstock cost	547	800	601
Biodiesel production cost	137	196	150
Total	684	996	751

Source: Yusuf et al. (2011)

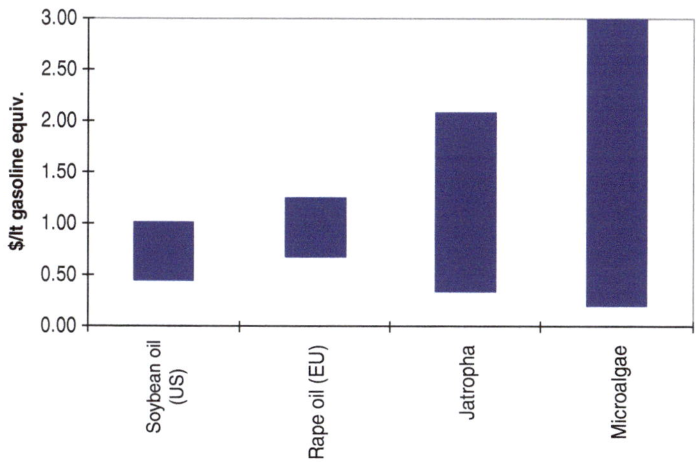

Fig. 3.4 Cost of various feedstocks for first and second generation biofuels. *Source*: Carriquiry et al. (2011)

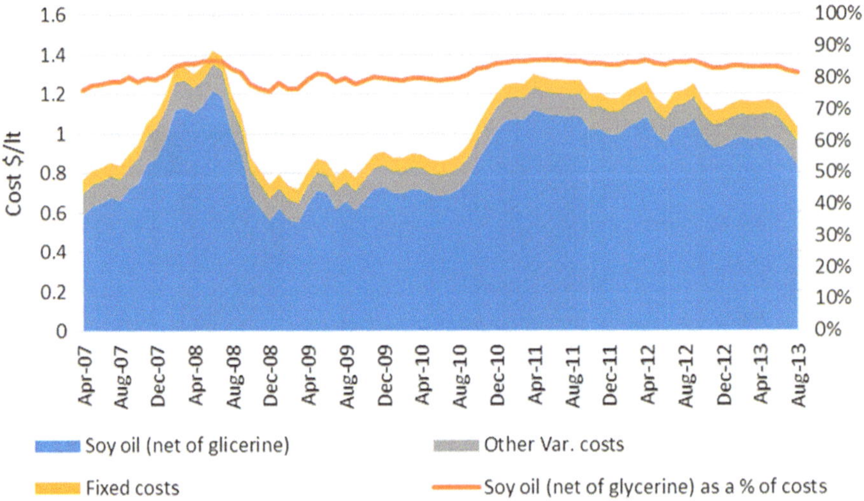

Fig. 3.5 Monthly costs of soybean oil-based biodiesel production for representative plants in Iowa, US. *Source*: http://www.extension.iastate.edu/agdm/reoutlook.html

The role of feedstock in the total production costs of biodiesel is illustrated in Fig. 3.5. For the period between April of 2007 and August of 2013, feedstock costs accounted for 76–86 % of the total costs of biodiesel production. If fixed costs are excluded, feedstocks would represent 84–91 % of variable costs of production over the period considered. Therefore, any assessment of biodiesel production costs is sensitive to location and time period considered.

3.3 Cost of Second Generation Biofuels

Most assessments of costs of producing cellulosic biofuels are based on ex-ante techo-economic analysis because of the absence of large-scale plants for commercial production and the proprietary nature of the information generated coming from those companies involved in some commercial production. Based on the available ex-ante assessments, the costs of second generation biofuels are much higher compared to (1) fossil fuels they replace and (2) first generation biofuels. Table 3.3 summarized estimations of the production costs of second generation biofuels reported in the existing literature. There is a wide variation in the cost estimates for reasons highlighted earlier, moreover no clear trend of costs exhibits over time. Table 3.4 presents operating costs and investment need for second generation biofuels that use thermo-chemical processes to derive liquid biofuels from biomass. The costs

Table 3.3 Estimated costs of production of different cellulosic biofuels

Study	Feedstock	Biofuel	Total cost ($/L gasoline equivalent)[a]
McAloon et al. (2000)	Corn stover	Ethanol	0.95
Solomon et al. (2007)	Switchgrass or wood	Ethanol	0.95
Sassner et al. 2008	Salix (willow)	Ethanol	0.90–1.09
	Spruce	Ethanol	0.82–0.87
	Corn stover	Ethanol	0.84–1.08
Frederick et al. (2008)	Yellow poplar	Ethanol	0.63
	Loblolly pine	Ethanol	0.71[b]
	Loblolly pine	Ethanol	1.03[b]
Wright et al. (2010)	Corn stover	Hydrocarbons	0.58
Kazi et al. (2010)	Corn stover	Ethanol	1.41–2.38[c]
Swanson et al. (2010)	Corn stover	Hydrocarbons	1.10–1.37
Brown et al. (2013)	Corn stover	Hydrocarbons	0.68
Haque and Epplin (2012)	Switchgrass	Ethanol	0.66–1.08

Source: HLPE (2013)
[a]Inflation adjusted to 2012
[b]Two different pre-treatments of the biomass
[c]The upper limit refers to the pioneer plant, whereas the lower limit refers to the *n*-th plant (i.e. benefiting from previous plant experience)

Table 3.4 Production costs of second generation biofuels using thermo-chemical process

Author (year)	Fuel cost (Eur/GJ)	Fuel costs standardized at 400 MW thermal (Eur 2011/GJ)	Investment (million 2011 Euros at 400 MW thermal)
Tijmensen et al. (2002)	19–25	19.31–24.7	781–906
Hamelinck (2004)	14–17.5	14–17.5	532–665
DENA (2006)	34.3	34.3	583
ECN (2006)	17.5–22.5	17.5–22.5	1052
RENEW (2006)	24.5–63.1	35.8–63.1	372–516
Festel (2007)	27.3	27.3	1224
McKeoug and Kurkela (2008)	15.8	15.8	508
van Vliet et al. (2009)	18–29	18–29	598–671
Swanson et al. (2010)	26.2–32.6	35.9–44.7	603–734
Tock et al. (2010)	20	20	311
Hohwiller (2011)	31.3–37	31.3–37	695–790
Departe (2010)	26	26	729
Sues (2011)	40	40	469
Liu et al. (2011)	19.6–21.3	30.1–32.8	436–444
Haarlemner et al. (2012)	38.7–38.7	38–38.7	834–949

Source: Elaborated based on Table 1 in Haarlemner et al. (2012)

Table 3.5 Production cost breakdown of second generation biofuels ($/L)[a]

	Sassner et al. (2008)			McAloon et al. (2000)	Solomon et al. (2007)	Frederick et al (2008)		
	Salix (willow)	Spruce	Corn stover	Corn stover	Switchgrass or Wood	Yellow Poplar	Loblolly Pine (1)[b]	Loblolly Pine (2)
Feedstock	0.23–0.28	0.21–0.23	0.21–0.28	0.19	0.2	0.15	0.23	0.27
Other costs	0.19–0.26	0.17–0.19	0.18–0.26	0.2	0.22	0.12	0.11	0.07
Co-products	−0.25	−0.22	−0.25	−0.02	−0.04	−0.02	−0.02	−0.11
Total operating costs	0.32–0.37	0.28–0.3	0.3–0.37	0.36	0.38	0.25	0.32	0.23
Capital costs	0.25–0.31	0.24–0.25	0.23–0.31	0.24	0.22	0.14	0.12	0.42
Total costs	0.57–0.69	0.52–0.55	0.53–0.68	0.6	0.6	0.39	0.44	0.65

[a]Inflation adjusted to 2008
[b]Two different pretreatments of the biomass are considered here

presented in Tables 3.3 and 3.4 are sensitive to various factors such as conversion efficiencies of the processes, the number of production pathways, assumed capital, and operational costs (Brown and Brown 2013; Stephen et al. 2012).

Table 3.5 presents the breakdown of production costs of second generation biofuels into various components, such as feedstock costs, other operating costs, and capital costs. In contrast to the first generation biofuels where feedstock costs account for more than two-thirds of the total production costs; feedstock costs account for around half (or lower) of the total production costs in the case of second generation biofuels.

The technical potential for biofuel production depends critically on both the amount of land that would be available, and the biomass productivity. While biofuels could technically make significant contributions to the global energy supply, their market potential is likely to be more limited due to the amount of feedstocks that can be economically produced and harvested as well as their costs relative to those of liquid fossil fuels (Carriquiry et al. 2011). The next section provides an overview of the costs of the major feedstocks being considered.

3.3.1 Ethanol from agricultural residues

Production processes for herbaceous energy crops are not standardized and are also subject to regional and climate conditions. Cost assessments are also not standardized. Table 3.6 compares the costs found for some herbaceous crops. As costs rise for sugar and starch foods, hopes for advanced biofuel technologies abound. Lignocelluloses are already considered a carbon neutral trade for energy (Searchinger et al. 2008). There is no additional requirement of land, land-use changes, or agricultural inputs. Crop residue extraction may, on the one hand, be beneficial for controlling soil temperatures, pests, and diseases (Andrews 2006). On the other hand, it is argued that residue removal may alter soil properties, soil productivity, water conservation, and carbon sequestration (Blanco-Canqui and Lal 2009).

Table 3.6 Costs of second generation biofuels produced from residues of herbaceous crops

	Estimated cost		
Feedstock	$/t	$/L ethanol	Source
Corn stover	54–65	0.188–0.224	Petrolia (2008)
	38–43	0.131–0.148	Petrolia (2006)
	76	0.262	Tokgoz et al. (2007)
	54.67	0.189	Frederick et al. (2008)
Winter wheat, continuous	20.16–28.04	0.070–0.097	Gallagher et al. (2003)
Winter wheat, fallow	38.18	0.132	Gallagher et al. (2003)
Spring wheat, continuous	24.17	0.083	Gallagher et al. (2003)
Sorghum	21.25–23.16	0.079–0.086	Gallagher et al. (2003)
Barley	21.78	0.07	Gallagher et al. (2003)
Oats	23.18	0.089	Gallagher et al. (2003)
Rice	25.21	0.09	Gallagher et al. (2003)

3.3.2 Forest Residues

Another source of second-generation biofuels, "forest residues" encompasses organic materials left from logging harvests, fuel wood extraction, and mill residues from primary and secondary wood processing (Perlack et al. 2005). While estimating a current extraction potential of 65 gal for fuel per ton of feedstock, the National Renewable Energy Laboratory (NREL) had set goals for 94 gal fuel yield per ton of feedstock for 2020. In assessment, however, the conversion rate of forest residues is lower in actuality than in theory. Gaps between theoretical conversion rates are expected to diminish over time. The greatest impediment to realizing theoretical yield is due to inaccessibility. Gathering residue from accessible areas would aggrandize costs to existing wood logging and collection. Furthermore, recovery of forest residue also draws environmental concern for sequestration, soil restoration, and forest regeneration cycles. Table 3.7 compares estimated costs of ethanol produced from forest residues.

3.3.3 Energy Crops

The biofuel industry will rely on a variety of feedstock: from herbaceous to woody, from arable to residue, from temporal to perennial. Comparative advantages usually dictate what regional inputs are used for production to fuel. However, an overarching comparison might be made for some dedicated feedstocks. Dedicated feedstocks are so chosen for being relatively less input-intensive, easier on soil, and higher yield altogether. In this section, we briefly mention the dedicated energy crops that have received the most attention in the literature as feedstocks for cellulosic biofuels. Estimates of the costs of production of selected herbaceous crops are presented in Table 3.8.

Table 3.7 Costs of second generation biofuels produced from forest residues

Feedstock	Estimated cost $/t	$/L ethanol	Source
Hardwood primary mill residue	33.9	0.113	NREL (1998)
Softwood primary mill residue	34.6	0.115	NREL (1998)
Hardwood secondary mill residue	30.5	0.102	NREL (1998)
Softwood secondary mill residue	30.4	0.102	NREL (1998)
Primary forest fuel (residues)	27	0.09	Junginger et al. (2005)[a]
Poplar	110–132	0.365–0.438	Manzone et al. (2009)[b]
Yellow poplar	48.1	0.16	Frederick et al. (2008)
Loblolly pine	67.0–71.5	0.22–0.24	Frederick et al. (2008)
Slash pine	43.18[c]	0.306–0.631	Nesbit et al. (2011)
Loblolly pine	69.4	0.341	Gonzalez et al. (2012)
Natural hardwood	71.0	0.362	Gonzalez et al. (2012)

Source: Elaborated by the authors

[a]Originally reported in 2002 €/Gj, converted using 21.1 Mj/L of ethanol (LHV) a yield of 300 L/t of forest residues, an exchange rate of 1.08 €/$, and updated to 2008 dollars using the GDP deflator (multiplied by 1.175)

[b]Original in €/t, converted with an exchange rate of 0.68 €/$ and 300 L of ethanol per ton of biomass

Table 3.8 Estimated costs of herbaceous energy crops delivered to a bio-refinery

Feedstock	Estimated cost[a] $/t	$/L ethanol	Source
Alfalfa	77–90	0.257–0.3	Vadas et al. (2008)
Miscanthus	51	0.169	Aravindhakshan et al. (2010)
Grassy biomass	27–59	0.090–0.197	Mapemba et al. (2007)
Reed canarygrass	65–98	0.217–0.327	Hallam et al. (2001)
Switchgrass	46–88[b]	0.153–0.293[b]	Perrin et al. (2008)
Switchgrass	56–60	0.187–0.200	Vadas et al. (2008)
Switchgrass	50-67	0.167–0.222	Epplin et al. (2007)
Switchgrass	92–121	0.307–0.402	Babcock et al. (2007)
Switchgrass	116	0.387	Duffy (2007)
Switchgrass	29	0.097	Pimentel and Patzek (2005)
Switchgrass	55–60	0.182–0.199	Haque and Epplin (2010)
Switchgrass	43	0.144	Aravindhakshan et al. (2010)

Source: Carriquiry et al. (2011)

[a]Inflation adjusted to 2008

[b]Does not include transportation costs to the biorefinery

Among herbaceous biofuel crops:

- Switchgrass: the Biofuels Feedstock Development Program has identified switchgrass as the most promising bioenergy crop (ORNL 2008) out of 34 herbaceous species. Switchgrass is touted based on its relatively low water and input requirements, environmental benefits, and ability to subsist in lower quality lands (Keshwani and Cheng 2009). Switchgrass naturally grows in North America, over a variety of soils so long as water retention is met.

- Miscanthus: native to Asia, miscanthus, is resilient to cold temperatures and low-nitrogen soils. The limitations of this crop involve its narrow genotype (making the crop less adaptable), seasonality, and high costs and long time to establish (requiring wide distribution of rhizome cuttings).
- Reed canarygrass: is typically used for forage or hay. Although reed canarygrass is suited to temperate climates and erosive conditions, once established in wetlands, it can become an invasive species.
- Alfalfa: an important forage crop used throughout the world. Once adjusted to a habitat, alfalfa is a high-nutrition animal feed. Reproduction of alfalfa requires pollination by well-adapted bees.

Woody feedstock crops generally have higher yield and greater versatility as solid or liquid energy (Cheng and Timilsina 2011). Among the faster growing tree species, include:

- Polar: a genus of 25–35 species of flowering plants. Poplar is attractive for its high-energy input/output ratio, carbon mitigation potential, and growth rate next to other woody plants. Potential downsides to harvesting poplars for biomass include wind erosion in stripped fields.
- Willow: a genus of about 400 species of deciduous trees and shrubs (Mabberley 1997), willows are highly cross-fertile and adaptable. While the willow serves many environmental benefits: soil erosion control, windbreak, soil reclamation, and shelterbelt, willow roots are aggressive spatially and tend to drain water resources. In Australia, willows have even become regarded as weeds.
- Eucalyptus: a genus of more than 700 species concentrated in Australia, eucalyptus can be an invasive species when transplanted. Its oil is multi-purposeful and is highly flammable – which can be a fire hazard in growth—but a highly efficient fuel.

3.4 Closing Remarks

The production costs of biofuels depend on type of feedstocks, geographic locations, exchange rates, and period for which the cost estimations are made. While Brazilians' sugarcane-based ethanol has traditionally been the lowest production costs, recent developments in markets, such as the expansion of domestic demand in Brazil, and increasing prices of sugar, have made US corn ethanol to be produced at similar costs. The relative competitiveness of these biofuels changes over time depending on relative feedstock prices and exchange rates. While costs of ethanol are closer to prices of gasoline, costs of biodiesel are much higher as compared to diesel due mainly to the elevated prices of feedstocks (vegetable oils and fats) which account for over 80 % of the total costs of production. For second-generation biofuels, the conversion cost is the key cost component, accounting for more than half of the total costs in most cases. It is therefore, the reduction in conversion costs through research and development is crucial to make second generation biofuels competitive to their first generation counterparts as well as fossil fuel they replace. An increase in feedstock yield does not only lower the overall cost of production but also minimizes the environmental footprint of biofuels.

References

Andrews, S. S. (2006). *Crop residue removal for biomass energy production: Effects on soils and recommendations* (White Paper). USDA-Natural Resource Conservation Service. Retrieved from http://soils.usda.gov/sqi/management/files/agforum_residue_white_paper.pdf

Aravindhakshan, S., Epplin, F., & Taliaferro, C. (2010). Economics of switchgrass and miscanthus relative to coal as feedstock for generating electricity. *Biomass and Bioenergy, 34*, 1375–1383.

Babcock, B. A., Gassman, P. H., Jha, M. J., & Kling, C. L. (2007). *Adoption subsidies and environmental impacts of alternative energy crops* (CARD Briefing Paper 07-BP-50). Ames, IA: Center for Agricultural and Rural Development, Iowa State University.

Blanco-Canqui, H., & Lal, R. (2009). Crop residue removal impacts on soil productivity and environmental quality. *Critical Reviews in Plant Sciences, 28*(3), 139–163.

Brown, T. R., & Brown, R. C. (2013). A review of cellulosic biofuel commercial-scale projects in the United States. *Biofuels, Bioproducts & Biorefining*. doi:10.1002/bbb.1387.

Brown, T. R., Thilakaratne, R., Brown, R. C., & Hu, G. (2013). Techno-economic analysis of biomass to transportation fuels and electricity via fast pyrolysis and hydroprocessing. *Fuels, 106*, 463–469.

Carriquiry, M. A., Du, X., & Timilsina, G. R. (2011). Second-generation biofuels: Economics and policies. *Energy Policy, 39*(7), 4222–4234.

Chen, X., & Khanna, M. (2012). Explaining the reductions in US corn ethanol processing costs: Testing competing hypothesis. *Energy Policy, 44*, 153–159.

Cheng, J. J., & Timilsina, G. R. (2011). Status and barriers of advanced biofuel technologies: A review. *Renewable Energy, 36*(12), 3541–3549.

Crago, C. L., Khanna, M., Barton, J., Giuliani, E., & Amaral, W. (2010). Competitiveness of Brazilian sugarcane ethanol compared to US corn ethanol. *Energy Policy, 38*, 7404–7415.

Departe, A. (2010). *Etude prospective sur la seconde generation de biocarburants, 2010/03*. Paris, France: DG Tresor. 2010.

Deutsche Energie-Agentur, G. H. (2006). *Biomass to liquid—BtL implementation report summary* (p. 18). Cologne, Germany: DENA.

Doornbosch, R., & Steenblik, R. (2007). *Biofuels: Is the cure worse than the disease?* Roundtable of Sustainable Development, OECD, Paris, France. Retrieved from http://www.oecd.org/sd-roundtable/39411732.pdf

Duffy, M. (2007). *Estimated costs for production, storage, and transportation of switchgrass*. Ames, IA: Iowa State University, University Extension. Retrieved July 2009 from http://www.econ.iastate.edu/research/webpapers/paper_12917.pdf

ECN. (2006). H. Boerrigter, economy of biomass-to-liquids (BTL) plants—An engineering assessment, ECN-C-06-019, ECN.

Epplin, F., Clark, C. D., Roberts, R. K., & Hwang, S. (2007). Challenges to the development of a dedicated energy crop. *American Journal of Agricultural Economics, 89*, 1296–1302.

Festel, G. (2007). *Biofuels—Which one is the most economic one?* (p. 23) Scientific Forum at ILMAC, Basel.

Frederick, W. J., Lien, S. J., Courchene, C. E., DeMartini, N. A., Ragauskas, A. J., & Iisa, K. (2008). Co-production of ethanol and cellulose fiber from southern pine: A technical and economic assessment. *Biomass and Bioenergy, 32*, 1293–1302.

Gallagher, P., Dikeman, M., Fritz, J., Wailes, E., Gauther, W., & Shapouri, H. (2003). Biomass from crop residues: Cost and supply estimates. *USDA Agricultural Economic Report, 819*.

Gonzalez, R., Daystar, J., Jett, M., Treasure, T., Jameel, H., Venditti, R., et al. (2012). Economics of cellulosic ethanol production in a thermochemical pathway for softwood, hardwood, corn stover and switchgrass. *Fuel Processing Technology, 94*, 113–122.

Haarlemner, G., Boissonnet, G., Imbach, J., Setier, P. A., & Peduzzi, E. (2012). Second generation BtL type biofuels—A production cost analysis. *Energy and Environmental Sciences, 5*, 8445–8456.

Hallam, A., Anderson, I. C., & Buxton, D. R. (2001). Comparative economic analysis of perennial, annual, and intercrops for biomass production. *Biomass and Bioenergy, 21*, 407–424.

Hamelinck, C. N. (2004). *Outlook for advanced biofuels*. Utrecht, The Netherlands: University of Utrecht. 2004.
Haque, M., & Epplin, F. (2010, July 25–27). *Switchgrass to ethanol: A field to fuel approach*. Paper Presented at 2010 AAEA Annual Meeting, Denver, CO.
Haque, M., & Epplin, F. M. (2012). Cost to produce switchgrass and cost to produce ethanol from switchgrass for several levels of biorefinery investment cost and biomass to ethanol conversion rates. *Biomass and Bioenergy, 46*, 517–530.
Hettinga, W. G., Junginger, H. M., Dekker, S. C., Hoogwijk, M., McAloon, A. J., & Hicks, K. B. (2009). Understanding the reductions in US corn ethanol production costs: An experience curve approach. *Energy Policy, 37*, 190–203.
HLPE. (2013). *Biofuels and food security*. A report by the high level panel of experts on food security and nutrition of the committee on world food security, Rome.
Hohwiller, C. (2011). *La production de carburants liquids par thermoconversion de biomasse lignocelluloqique: Evaluation pour le systeme energetique francais future*. Paris, France: ParisTech.
Junginger, M., Faaij, A., Bjorheden, R., & Turkenburg, W. C. (2005). Technological learning and cost reductions in the wood fuel supply chains in Sweden. *Biomass and Bioenergy, 29*(6), 399–418.
Kazi, F. K., Fortman, J. A., Anex, R. P., Hsu, D. D., Aden, A., & Dutta, A. (2010). Techno-economic comparison of process technologies for biochemical ethanol production from corn stover. *Fuel, 89*(1), S20–S28.
Keshwani, D., & Cheng, J. (2009). Switchgrass for bioethanol and other value-added applications: A review. *Bioresource Technology, 100*, 1515–1523.
Klein-Marcuschamer, D., Oleskowicz-Popiel, P., Simmons, B. A., & Blanch, H. W. (2012). The challenge of enzyme cost in the production of lignocellulosic biofuels. *Biotechnology and Bioengineering, 4*, 1083–1087.
Liu, G., Larson, E. D., Williams, R. H., Kreutz, T. G., & Guo, X. (2011). Making Fischer–Tropsch fuels and electricity from coal and biomass: Performance and cost analysis. *Energy & Fuels, 25*, 415–437.
Mabberley, D. J. (1997). *The plant book* (2nd ed.). Cambridge, England: Cambridge University Press.
Manzone, M., Airold, G., & Balsari, P. (2009). Energetic and economic evaluation of a poplar cultivation for the biomass production in Italy. *Biomass & Bioenergy, 33*, 1258–1264.
Mapemba, L., Eppin, F. M., Taliaferro, C. M., & Huhnke, R. L. (2007). Biorefinery feedstock production on conservation reserve program land. *Review of Agricultural Economics, 29*, 227–246.
McAloon, A., Taylor, F., Yee, W., Ibsen, K., & Wooley, R. (2000). Determining the cost of producing ethanol from corn starch and lignocellulosic feedstocks. *NERL Technical Report, 580–28893*.
McKeough, P., & Kurkela, E. (2008). *Process evaluations and design studies in the UCG project 2004-2007*. Research notes 2434 (p. 49). Espoo, Finland: VTT Tiedotteita.
National Renewable Energy Laboratory (NREL). (1998, September). *Lignocellulosic feedstock resource assessment*. TP-580-24189
Nesbit, T. S., Alavalapati, J. R., Dwivedi, P., & Marinescu, M. V. (2011). Economics of ethanol production using feedstock from slash pine (Pinus elliottii) plantations in the southern United States. *Southern Journal of Applied Forestry, 35*(2), 61–66.
Oak Ridge National Laboratory (ORNL). (2008). *Exploring potential US switchgrass production for lignocellulosic ethanol*. ORDL/TM-2007/183.
Perlack, R., Wright, L., Turhollow, A., Graham, R., Stokes, B., & D. Erbach. (2005, April). *Biomass as feedstock for a bioenergy and bioproducts industry: The technical feasibility of a billion-ton annual supply*. Department of Energy/GO-102005-2135.
Perrin, R. K., Vogel, M. S., & Mitchell, R. (2008). Farm-scale production cost of switchgrass for biomass. *Bioenergy Resources, 1*, 91–97.
Petrolia, D. R. (2006, July 23–26). *Ethanol from biomass: Economic and environmental potential of converting corn stover and hardwood forest residue in Minnesota*. Paper presented at the American Agricultural Economic Association Annual Meeting, Long Beach, CA.
Petrolia, D. R. (2008). The economics of harvesting and transporting corn stover for conversion to fuel ethanol: A case study for Minnesota. *Biomass and Bioenergy, 32*, 603–612.

Pimentel, D., & Patzek, T. W. (2005). Ethanol production using corn, switchgrass, and wood; biodiesel production using soybean and sunflower. *Natural Resources Research, 14*, 65–76.
RENEW; Vogel, A., Brauer, S., Muller-Langer, F., & Thran D. (2006). *RENEW project—Conversion cost calculation deliverable D5.3.7.* Renewable fuels for advanced powertrains, SES6-CT-2003-502705, Institute for Energy and Environment, Leipzig, Germany.
Sassner, P., Galbe, M., & Zacchi, G. (2008). Techno-economic evaluation of bioethanol production from three different lignocellulosic materials. *Biomass and Bioenergy, 32*(2008), 422–430.
Searchinger, T., Heimlich, R., Houghton, R., Dong, F., Elobeid, A., Fabiosa, J., et al. (2008). Use of US croplands for biofuels increases greenhouse gases through emissions from land use change. *Science, 319*(5867), 1157–1268.
Solomon, B., Barnes, J., & Halvorsen, K. (2007). Grain and cellulosic ethanol: History, economics, and energy policy. *Biomass and Bioenergy, 31*, 416–425.
Stephen, J. D., Mabee, W. E., & Saddler, J. N. (2012). Will second-generation ethanol be able to compete with first-generation ethanol? Opportunities for cost reduction. *Biofuels, Bioproducts and Biorefining, 6*, 159–176.
Sues, A. (2011). *Are European bioenergy targets achievable?* Eindhoven, The Netherlands: Eindhoven University of Technology.
Swanson, R. M. A., Platon, A., Satrio, J. A., & Brown, R. C. (2010). Technoeconomic analysis of biomass-to-liquids production based on gasification. *Fuel, 89*, S11–S19.
Tijmensen, M. J. A., Faaij, A. P. C., Hamelinck, C. N., & Hardeveld, M. R. M. (2002). Exploration of the possibilities of Fischer–Tropsch liquids and power via biomass gasification. *Biomass and Bioenergy, 23*, 129–152.
Tock, L., Gassner, M., & Marechal, F. (2010). Thermochemical production of liquid fuels from biomass: Thermo-economic modelling, process design and process integration analysis. *Biomass and Bioenergy, 34*, 1838–1854.
Tokgoz, S., Elobeid, A., Fabiosa, J., Hayes, D. J., Babcock, B. A., Yu, T., et al. (2007). *Emerging biofuels: Outlook of the effects on US grain, oilseeds, and livestock markets Staff Report 07-SR-101.* Ames, IA: Center for Agricultural and Rural Development, Iowa State University.
Vadas, P. A., Barnett, K. H., & Undersander, D. J. (2008). Economics and energy of ethanol production from alfalfa, corn, and switchgrass in the Upper Midwest, USA. *Bioenergy Resources, 1*, 44–55.
van den Wall Bake, J. D., Junginger, M., Faaij, A., Poot, T., & Walter, A. (2009). Explaining the experience curve: Cost reductions of Brazilian ethanol from sugarcane. *Biomass and Bioenergy, 33*, 644–658.
van Vliet, O. P. R., Faaij, A. P. C., & Turkenburg, W. C. (2009). Fischer–Tropsch diesel production in a well-to-wheel perspective: A carbon, energy flow and cost analysis. *Energy Conversion and Management, 50*, 855–876.
World Watch Institute (WWI). (2007). *Biofuels for transport: Global potential and implications for sustainable energy and agriculture.* Sterling, VA: Earthscan.
Wright, M. M., Daugaard, D. E., Satrio, J. A., & Brown, R. C. (2010). Techno-economic analysis of biomass fast pyrolysis to transportation fuels. *Fuel, 89*(1), S2–S10.
Yusuf, N., Kamarudin, S. K., & Yaakub, Z. (2011). Overview on the current trends in biodiesel production. *Energy Conversion and Management, 52*, 2741–2751.

Chapter 4
Impacts of Biofuels on Food Prices

Gal Hochman, Deepak Rajagopal, Govinda R. Timilsina, and David Zilberman

4.1 Introduction

For the first time since the Green Revolution, food and fuel commodity prices began to rise in 2001 and reached peak levels in 2008 (FAO 2008; Peters et al. 2009; Trostle 2008a, b). As primary food commodities (e.g., corn, wheat) doubled in price, the biofuel industry expanded manifold within this same period (e.g., ethanol from corn and sugarcane doubled to 65 billion liters; biodiesel from soybean, oil palm, and rapeseed reached 12 billion liters or six times 2001 levels (Martinot and Sawin 2009). Popular opinion has linked biofuel production to the shock in food prices in 2008. Yet, much of the biofuel demand by the US and EU was driven by government mandates and subsidies, which aim to reduce the demand for oil and increase demand for agricultural goods (Hochman et al. 2010a, b, c). In high-income households and countries, a smaller percentage of crop prices is reflected in the final food price (due to food processing, packaging services). In low-income households and countries, however, crop prices have much larger share of the final food price.

G. Hochman (✉)
Rutgers University, 55 Dudley Rd., Cook Campus, Rutgers University, New Brunswick 08901, NJ, USA
e-mail: hochman@aesop.rutgers.edu

D. Rajagopal
Institute of the Environment and Sustainability, UCLA, Los Angeles, CA, USA
e-mail: rdeepak@ioes.ucla.edu

G.R. Timilsina
The World Bank, 1818 H Street, N.W., Washington, DC 20433, USA
e-mail: gtimilsina@worldbank.org

D. Zilberman
Department of Agricultural and Resource Economics, College of Natural Resources, University of California, Berkeley, Giannini Hall 206, Berkeley, CA 94720-3310, USA
e-mail: zilber11@berkeley.edu

G.R. Timilsina and D. Zilberman (eds.), *The Impacts of Biofuels on the Economy, Environment, and Poverty*, Natural Resource Management and Policy 41,
DOI 10.1007/978-1-4939-0518-8_4, © Springer Science+Business Media New York 2014

Hochman et al. (2011) identify and quantify the major factors in food commodity prices and also report the importance of the ability to store certain feedstock and of inventory relative to the consumption levels. It is argued that price shock for food commodities in 2008 was partly caused by the declining stock-to-use ratio since 1985 (Trostle 2008a, b).

Historical lows in inventory caused societies to experience greater price sensitivity. Other notable factors in the food crisis included global economic growth, population growth, energy price inflation, exchange rate fluctuations, adverse weather, and trade policy. Economic growth created new demand for luxury foods and meats (which heighten production costs) by rising income households. Energy price inflation motivated many farmers to convert agricultural plots to biofuel feedstock plantations. In 2008, the US dollar depreciated relative to major world currencies (Abbott et al. 2008)—thus raising commodity prices for many nations in trade with the United States and inciting currency speculation during the exchange rate flux (Rosegrant 2008). Adverse weather patterns experienced in key grain-producing regions caused high production costs and shortage of crop output. Many of these factors prompted governments to ensure their own food supply by stalling grain trades, consequently straining regions relying on food imports (Timmer 2008). In addition, cumulative underinvestment in agricultural research and technology stagnated agriculture's productivity growth (Schnepf 2004).

4.2 Prior Research

Economic assessments of the food crisis have been categorized as based on partial or general equilibrium models. Partial equilibrium models—e.g., IMPACT, AGLINK/COSMO, FAPRI, and FASOM—utilize supply and demand equations to represent the economic behavior within select markets (Alston et al. 2009). Disadvantages arise from partial equilibrium analyses, because it is not comprehensive enough to gauge restraints on budget. Computable general equilibrium (CGE) models—e.g., GTAP, LINKAGE, and USAGE—could correct some disadvantages, but would require much more robust data and complexity. CGE is a numerical technique based on Walrasian theory that was formalized by Arrow and Debrew to model supply, demand, and prices across a set of markets. Tables 4.1 and 4.2 summarize a range of studies that estimated commodity price effects.

As biofuels are increasingly used to substitute crude oil for energy uses, many have studied the impact that feedstock prices have on the market (Zilberman et al. 2012). The food crisis of 2007–08 has given further cause for research—wherein the most pessimistic of estimates attributed 75 % of the food price shock to the biofuel industry (Mitchell 2008). The IMF and other bodies of scholars have suggested factors such as: depreciation of the dollar, global changes in production such as weather shocks, changes in patterns of food consumption, trade policies, government incentives, and the role of biofuels in commodity price increases. The USDA had reported greater

Table 4.1 Quantitative estimates of impact of biofuel on food commodity prices

Source	Estimate (%)	Commodity	Time period
Mitchell (2008)	75	Global food index	Jan 2002–Feb 2008
Rosegrant (2008)	39	Corn	2000–2007
	21–22	Rice and wheat	2000–2007
OECD-FAO (2008)	42	Coarse grains	2008–2017
	34	Vegetable oils	2008–2017
	24	Wheat	2008–2017
Collins et al. (2008)	25–60	Corn	2006–2008
	19–26	US retail food	2006–2008
Glauber (2008)	23–31	Commodities	Apr 2007–Apr 2008
	10	Global food index	Apr 2007–Apr 2008
	4–5	US retail food	Jan–Apr 2008
Lampe et al. (2006)	35	Corn	Mar 2007–Mar 2008
	3	Global food index	Mar 2007–Mar 2008
Rajagopal et al. (2009)	15–28	Global corn price	2007–2008
	10–20	Global soy price	2007–2008
Hochman et al. (2011)	20–30	Corn	2002–2007
	5–10	Soybeans	2002–2007
Fischer et al. (2009)	11–51	Coarse grains	2008
De Hoyos and Medvedev (2009)	6	Global food index	2005–2007

Table 4.2 Impacts of increased biofuel production on food prices

Study	Coverage and key assumptions	Key impacts of biofuels on food prices
Abbott et al. (2008)	Rise in corn price from about US$2–6 per bushel accompanying the rise in oil price from US$40 in 2004 to US$120 in 2008	US$1 of the US$4 increase in corn price (25 %) due to the fixed subsidy of 51-cents per gallon of ethanol
Baier et al. (2009)	24 months ending June 2008; historical crop price elasticities from academic literature; bivariate regression estimates of indirect effects	Global biofuel production growth responsible for 17, 14, and 100 % of the rises in corn, soybean, and sugar prices, respectively, and 12 % of the rise in the IMF's food price index
Banse et al. (2008)	2001–2010; Reference scenario without mandatory biofuel blending, 5.75 % mandatory blending scenario (in EU member states), 11.5 % mandatory blending scenario (in EU member states)	Price change under reference scenario, 5.75 % blending, and 11.5 % blending, respectively Cereals: –4.5, −1.75, +2.5 % Oilseeds: −1.5, +2, +8.5 % Sugar: –4, −1.5, +5.75 %
Collins (2008)	2006/2007–2008/2009; Two scenarios considered: (1) normal and (2) restricted, with price inelastic market demand and supply	Under the normal scenario, the increased ethanol production accounted for 30 % of the rise in corn price; Under the restricted scenario, ethanol accounted for 60 % of the expected increase in corn prices.

(continued)

Table 4.2 (continued)

Study	Coverage and key assumptions	Key impacts of biofuels on food prices
Fischer et al. (2009)	(1) Scenario based on the IEA's WEO 2008 projections (2) Variation of WEO 2008 scenario with delayed second gen biofuel deployment (3) Aggressive biofuel production target scenario (4) and variation of target scenario with accelerated second gen deployment	Increase in prices of wheat, rice, coarse grains, protein feed, other food, and non-food, respectively, compared to reference scenario: (1) +11, +4, +11, −19, +11, +2 % (2) +13, +5, +18, −21, +12, +2 % (3) +33, +14, +51, −38, +32, +6 % (4) +17, +8, +18, −29, +22, +4 %
Glauber (2008)	12 months ending April 2008	Increased US biofuel production accounted for 25 % of the rise in corn prices and 10 % of the rise in global food prices
IMF (2008)	Estimated range covers the plausible values for the price elasticity of demand	Range of 25–45 % for the share of the rise in corn prices attributable to ethanol production increase in the US
Lazear et al. (2008)	12 months ending March 2008	US ethanol production increase accounted for 20 % of the rise in corn prices
Lipsky (2008) and Johnson (2008)	2005–2007	Increased demand for world biofuels accounts for 70 % of the increase in corn prices
Mitchell (2008)	2002-mid-2008; ad hoc methodology: impact of movement in dollar and energy prices on food prices estimated, residual allocated to the effect of biofuels	70–75 % of the increase in food commodity prices was due to world biofuels and the related consequences of low grain stocks, large land-use shifts, speculative activity, and export bans
Rosegrant et al. (2008)	2000–2007; Scenario with actual increased biofuel demand compared to baseline scenario where biofuel demand grows according to historical rate from 1990 to 2000	Increased biofuel demand accounted for 30 % of the increase in weighted average grain prices, 39 % of the increase in real corn prices, 21 % of the increase in rice prices and 22 % of the rise in wheat prices

Source: Timilsina and Shrestha (2011)

demands for grains among world consumers, while production gains in technology had slowed in growth rate.

Macroscopically, food prices had been in the decline with the ongoing development of farming techniques from cropland rotation to the Green Revolution. Then, the increase of trade via globalization presented higher rates of capital flow that amplified energy demands. As oil prices spiked in the 1970s, the global market was challenged to deal with the scale of energy demand and the scale of supply alternatives (Graff et al. 2009). The alternative of biofuel will favor or even revive the agricultural sector. Baka and Roland-Holst (2009) argue that, in the case of Europe, biofuel production will reduce trade rivalries and heighten energy security. Agricultural biotechnology and its markets vary given regional natural resources, competitive advantages, industry set up costs, institutional regulations, and other economic considerations.

4.3 Historical Price Trends

For centuries, economic development has improved agricultural efficiency and allowed for resources to shift to sectors such as manufacturing. Inasmuch, the expansion of consumption and production—biased toward manufacturing and against agriculture—raised food prices (Sexton et al. 2009). From a partial equilibrium perspective, this mechanism would also heighten food demand and further boost food prices. An interesting case in point is China: 20 % of the world's population (the world's largest consumer and producer of food) produces agricultural products at only 20–33 % of the rate that it produces non-agricultural goods (Alston et al. 2009). On a more global scale, economic development has afforded higher population and income levels. This shifts the lifestyle and consumption habits of people on an exceptionally larger scale, considering developing countries. The historical trends of production, consumption, inventory, and prices of major grains—corn, wheat, rice, soybeans, rapeseed, oil palm (see Fig. 4.1) show that crop prices have been countercyclical to their inventory levels. Consumption levels of wheat have decreased since 2004, while consumption levels of coarse grains and rice have increased. The higher demand for coarse grains mostly comes from US demand for corn for ethanol production. The higher demand for rice is concentrated in Asian countries, which have increased their consumption levels from 61.5 to 85.9 kg/capita. Rice crops are characteristically produced by nations for domestic consumers, under segmented and protected markets.

It is notable that, outside the agricultural industry, other commodity prices were also on the rise (e.g., minerals, metals, energy). Between January 2002 and July 2008, the IMF price indices showed that food prices rose 130 % and crude oil prices rose 590 %, contrasted against 330 % rise in commodity prices in general. The impact of biofuel demand on food prices manifests in two ways: allocation of land (for which biofuel feedstock compete with food products, thus increasing aggregate demand for agricultural commodities) and the level of energy prices (which affects production costs and output level of agricultural commodities).

When comparing allocation rate of corn, soybean, and rapeseed crops toward biofuel production, rapeseed has the highest share of its total supply allocated to biofuel, thus signifying that biofuel has become an important factor for the increasing demand and prices of rapeseed. While corn and soy also experienced rising biofuel allocation per total supply, biofuel appears less important a demand to affect corn and rice prices.

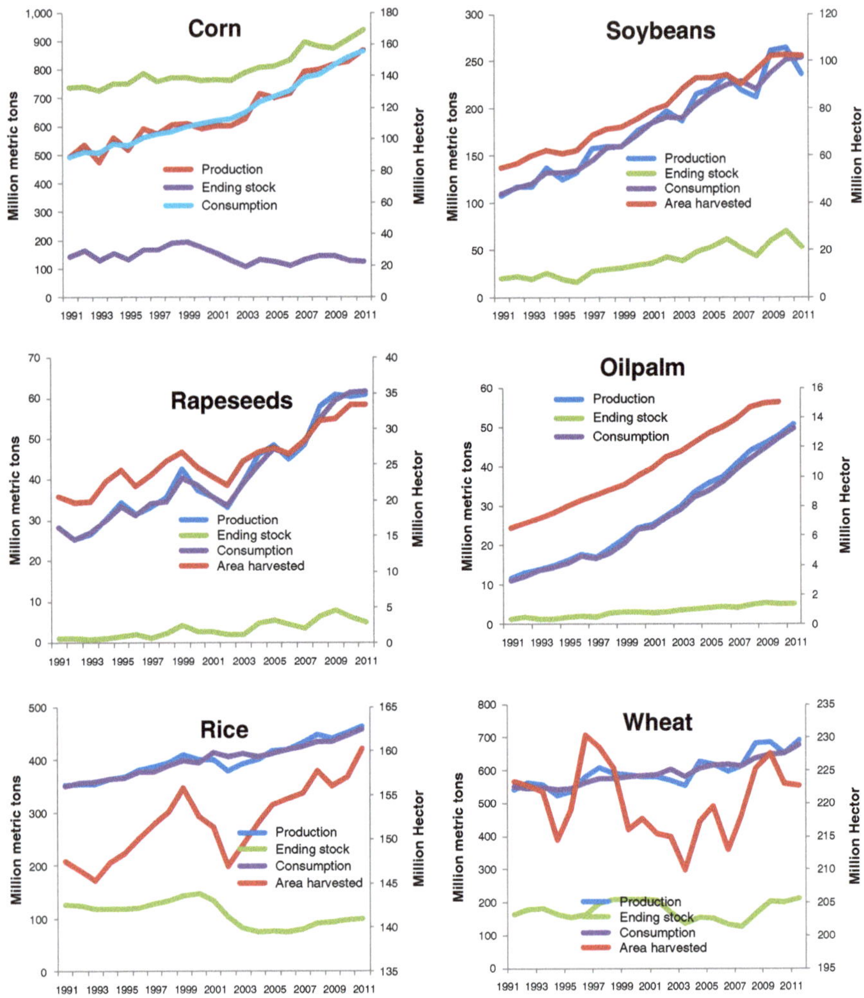

Fig. 4.1 Historical trends of prices, production, consumption, and inventory of course grains and oil crops

The food price shock was not instantaneous. On the demand side, consumption of agricultural products was rising across the world. On the supply side, production technologies were making less gains and growth had been sluggish. Agricultural output in developing nations had been almost half of their GDP growth for the past two decades. During this time period in which demand outpaced supply, stockpiles of grain commodities diminished with use. In fact, stock-to-use ratios declined by more than 50 % since 1985, making regional markets more sensitive to changes in grain prices. Depicted in Fig. 4.2, the observed correlation between price and inventory is graphed for major grain crops. Stock-to-use of the world's grain and oilseeds were recorded at 35 % in 1985 and at less than 15 % in 2005.

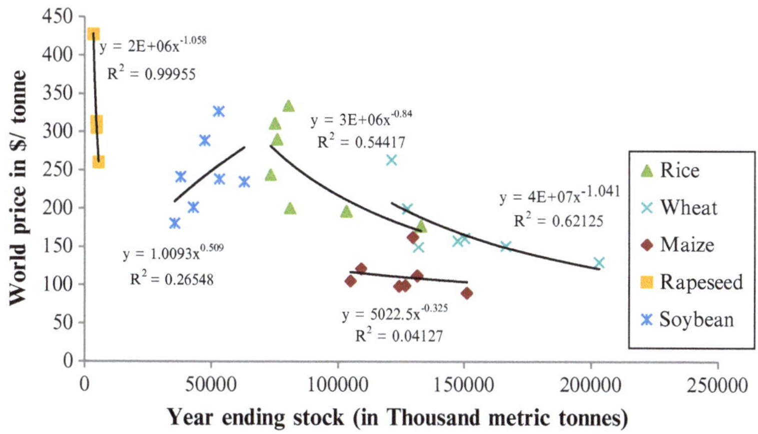

Fig. 4.2 The observed correlation between price and inventory

When modeling the adjustments in global markets to handle new consumers and the volume of demand, Ivanic and Martin showed that real agricultural prices would continue to rise in the next 40 years, requiring higher income, supply, and demand (Ivanic and Martin 2010). As diagrammed, the current scenario of demand expansion (due to the biofuel industry) results in a shortage of agricultural supply. The balance of supply and demand gets a break if grains can be stored. Inventory levels can contribute an extra source of supply when demand calls for it.

Figure 4.1 also depicts world consumption of various coarse grains and oil crops. These illustrate the upward trend in global consumption from 2001 to 2010 for the various crops. From 2001 to 2010, demand grew, albeit at uneven rates across crop types. For example, corn demand grew by about 30 %, as rapeseed demand grew by almost 100 % from 2001 to 2007. Moreover, demand growth was not symmetric across regions. Whereas globally consumption of all crops increased with income (at the world level, income is positively correlated with consumption, and world income grew throughout the period investigated), in some regions consumption of certain crops decreased. For example, corn, rice, and wheat consumption in China went down by 12.7 %, 23.7 %, and 20.9 %, respectively, although global consumption increased by 24.0 %, 3.2 %, and 4.3 %, respectively. Although the global recession of 2008 hampered the food-commodity price inflation of 2007/2008, in 2009/2010 consumption returned to display an upward trend and so did food-commodity prices. Further, between 2011 and 2017, corn prices are expected to rise 14 % and soybean oil prices are expected to be more stable with only a 5 % rise (Zilberman et al. 2010). Timilsina and Shrestha argue that Ethanol production might be sustainable without support if gasoline prices remain above US$3/gal (Timilsina and Shrestha 2011). Then, even if ethanol production was doubled, corn prices would still remain at about US$4/bushel.

4.4 Methodology

Grain storability and its inventory level would soften a grain's price volatility on the market. With this intuition, the authors have developed a new model to capture more accurately biofuels' impact on the agricultural commodities market. Assuming that the demand function follows on historical data on prices and inventory levels, any anticipation of future inventory decisions affects current behavior and is constrained by the fact that one cannot borrow from future inventory (e.g., inventory cannot be negative) (Williams and Wright 1991). Graphical results of this demand function depict that when market demand exceeds the harvest of crops, prices will rise if stocks are too low. The inventory function would present a significant buffer to demand, thus suggesting that a model neglecting grain inventory would overestimate price effects of biofuel (Hochman et al. 2011).

4.4.1 Multi-market Analysis

For a multiple market analysis, methodology differs from the partial equilibrium and general equilibrium models. An important aim is to disaggregate markets in order to accurately portray, explain, and predict price impacts from policies regulating specific markets. Disaggregation of the "vertical structure" (the chain of production) would distinguish between supply interactions and the different end-uses—whether for food processing or energy production. Disaggregation of the "horizontal structure" (across different staple crops) assesses the input prices and the feedback effects between different markets. GTAP, FAPRI, and IFPRI models have utilized multi-market in various studies, though none has included grain inventory before.

Horizontal Structure: Since different staple crops compete for the same inputs (e.g., land, labor, resources), there is negative cross-price elasticity between different staple crops. For example, higher demand for Chinese soybeans may trigger shifts in resources from corn cultivation to soybean cultivation. Consequently, the strong growth rate of soybeans in China causes other competitive crops to lessen in production rate. A similar comparison can be made between agricultural production of biofuel to food. If the demand rises for corn-ethanol allocates more land to corn specifically for ethanol, then resources are detracted from food production from corn or other crops. Thus, biofuel also has a negative cross-price elasticity with respect to food. Yet, biofuel can sometimes replace one feedstock with another—thus, rendering different crops complementary and not substitutable goods. To properly disaggregate the horizontal structure, it is critical to determine the dominant forces between demand of agricultural inputs (e.g., land, energy) and end-use (e.g., food consumption, energy).

Vertical Structure: The vertical structure highlights the supply chain interactions. Production level would be determined by existing markets and introduction of new markets (e.g., ethanol, biodiesels). Differences in willingness to pay will incentivize suppliers to develop for one market over another. For example, farmers switched crops from food to biofuel feedstock and limited the sale to food markets. The vertical structure of production is likewise influenced by inputs (e.g., land, energy) and demand of end product (e.g., food, biofuel).

4.4.2 Numerical Model

Combining the horizontal and vertical structure intuition, the authors (Hochman et al. 2011) extend the empirical model for a single region, single crop to a multi-market with five major grain crops (corn, soybean, rapeseed, rice, wheat). They divide the world into seven major regions, namely, Argentina, Brazil, China, European Union (EU-27 countries), India, United States, and an aggregate that represents the rest of the world (ROW), and focus on the time period between the year 2001 and the year 2007.

To determine the level of inputs used, crop consumption minus the quantity of a co-product, which may be returned as an input. Biofuel from corn, soybean, and rapeseed is jointly produced along with a co-product that is itself a substitute for the raw grain or the oilseed. For instance, in the case of corn, 1 bushel (56 lb) of corn yields approximately 2.75 gal of ethanol and 18 lb of distiller grains, which is a substitute for corn grain. A fraction of the quantity of original crop used for biofuel is replaced in the form of co-product. Therefore, for these three crops, we compute an effective demand of the particular crop for biofuel, which equals the crop consumption for biofuel minus the quantity of a co-product. In the case of corn, the effective demand of corn is $0.68=(1-18/56)$ bushels per 2.75 gal of ethanol. Assume that biofuel production function is of Leontief (fixed-proportion) type. Further, when biofuel production is determined through a mandate, the derived crop demand for biofuel is simply a fixed proportion of the mandate.

With the exception of the demand for inventory, assume a linear structure for supply and demand. The linear structure generally serves as a good approximation for small disturbances or shocks. Crop demand for inventory is represented as a nonlinear function of price and follows Carter et al. (2008). For details regarding the calibration of the numerical model and the calculations made for various shocks, see Hochman et al. (2011).

4.4.3 Numerical Scenarios

Given the cumulative change in a variable with respect to the year 2001, use the market-clearing condition to derive a counterfactual equilibrium world price for each crop for the various shocks for each year. Repeating this process for four different alternative scenarios which either differ in the assumed range for elasticities used in calibration of supply and demand functions, or differ in the specification of the demand for food/feed (whether GDP per capita is explicitly represented in demand) or differ in parameters of the inventory demand function. Given the challenge of estimating a point estimate for the various elasticities, as well as the inventory parameters, we simulated these alternative scenarios to determine the robustness of our results.

The first scenario, which the authors henceforth refer to as the *baseline scenario*, is one in which used the range of price and income elasticities reported in the literature, namely, that mentioned in the USDA's database of elasticities and in the FAPRI database. Under this scenario, the parameters for the inventory demand function are those estimated using the specification of Carter et al (2008). As mentioned earlier, perform 100 simulations of this scenario for the various shocks for each crop and for each time period but report the mean value of these outcomes.

In the second scenario, the *inelastic scenario*, assume a narrower range for elasticities, which is on average more inelastic compared to the baseline scenario and follows Gardner (1987). This scenario further differs from the baseline in that we employ a demand specification that does not include income. The reason for excluding income is that some of the elasticities reported in the literature were based on models that did not include income.

Finally, to test the robustness of the inventory demand parameters, simulate a fourth scenario using Carter et al.'s (2008) estimates for the inventory demand function as opposed to the original estimates of this study. Note that Carter et al. (2008) estimated the inventory demand based on US data for 2006–2008, while the original results drew from world data in 2001–2008.

4.5 Results

The authors report two different price changes: First, reduction of commodity price if key variables would have stayed at 2001 levels, $\Delta P_{t,i}$. Technically, this is the percentage difference between the actual price in a given year and the counter-factual price for the same year, and secondly, the increase of the commodity price attributed to a change in one of the variables (income, biofuel mandate, exchange rate , energy prices) between 2001 and the specific year, $\Delta P_{t/2001,i}$, where $i \in \{$biofuel, income growth, energy prices, exchange rate$\}$. Technically, this is the percentage difference between counter-factual price for a given year and the price in 2001. The simulations compute $\Delta P_{t,i}$. The authors then compute $\Delta P_{t/2001,i}$ as follows: let ΔP_t^a denote the total percentage price change between the year t and year 2001; then,

$$\Delta P_{t/2001,i} = \Delta P_{t,i}\left(1 + \Delta P_t^a\right) / \Delta P_t^a \tag{4.1}$$

Total change in price from year t to year 2001 that is explained by this model equals the sum of $\Delta P_{t/2001,i}$ over all the shocks. The figures depict $\Delta P_{t,i}$—namely, the food commodity price reduction attributed to a shock that eliminates one of the factors that caused prices to change after 2001, whereas the tables show

$\Delta P_{t/2001,r}$—namely, the increase in commodity prices from 2001 attributed to one of the factors that caused prices to change after 2001. In both cases, the authors report the mean outcome of 100 simulations, where each trial draws upon a number from a range of plausible values (for price, income, and supply elasticities) and compute the counterfactual outcome. When presenting prices for different crops, the authors distinguish between two different specifications: one with inventory demand function and another without inventory demand. For each crop, the authors show the impact of these shocks one at a time.

The analysis includes five simulated scenarios for each of the five crops, namely, corn, soybeans, rapeseed, rice, and wheat. The baseline scenario's outcome is contrasted with alternative specifications to evaluate robustness of the relative and absolute value of the numerous shocks. The alternative scenarios illustrate the robustness of the results presented with respect to relative impact, but the absolute impact usually becomes larger as elasticities become smaller. Some but not all scenarios include an income term in the demand specification for food and feed. Introducing an income term reduces the biofuel impact. While for the first four scenarios the authors estimated an inventory demand function, for the fifth scenario, the authors relied on the parameters from Carter et al. as a constant. The estimated parameters suggest, on average, more elastic inventory demand, and thus less fluctuation in prices. The authors conclude this section by qualitatively discussing the role of trade policy and speculation and the role of inventory management for limiting the impact of future shocks.

4.5.1 The Baseline Scenario

The observed prices for the different crops are shown in Fig. 4.3. A clear upward trend, on average, emerges for all crops, albeit some prices increase more than others. Whereas the price of corn and soybeans increased from 2002 to 2006 by about 63 %, the price of wheat increased by more than 74 %. Furthermore, while some crops like rice and wheat experienced an upward trend throughout the period, others such as soybeans declined in 2005 and 2006 only to increase by 39 % in 2007. Inventory theory predicts that prices decline when inventory accumulates and vice versa. The data confirm these predictions, except for soybeans, and show similar trends for stock-to-use ratio (see Fig. 4.3). If, however, dropping 2007 (a year where soybean prices spiked), then such a pattern is also observed for soybeans.

Inventory serves as a buffer and affects prices as long as inventory levels are sufficiently large. However, as these levels become small, prices become more volatile and sensitive to the numerous specific factors affecting crop prices. Less fluctuation is observed if inventory demand is explicitly added to the analysis. The aggregate demand curve becomes much more elastic for large inventory levels, and thus predicts less price volatility (Fig. 4.4).

The annual increase in corn and soybean prices is largest toward the end of the sample period (i.e., between 2006 and 2007). One explanation for the observed

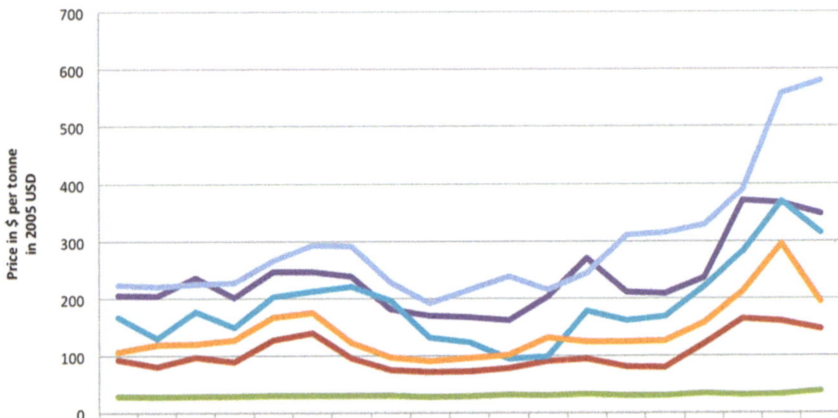

	1991	1992	1993	1994	1995	1996	1997	1998	1999	2000	2001	2002	2003	2004	2005	2006	2007	2008	2009
Corn	93	81	98	89	128	140	96	76	72	73	78	91	95	81	79	120	165	160	146
Sugar	29	28	29	29	30	31	31	30	28	29	32	31	33	31	31	34	32	33	38
Soybeans	205	204	235	201	247	247	238	181	170	167	161	203	270	211	208	236	371	366	347
Rice	167	130	176	149	202	212	221	196	131	124	94	99	178	162	169	220	282	370	315
Wheat	107	119	120	127	167	175	124	97	91	96	102	131	125	125	126	157	212	295	195
Rapeseeds	223	220	225	227	265	293	291	227	191	215	238	215	243	311	315	328	390	558	580

Fig. 4.3 Average annual crop prices (in 2005 USD per ton)

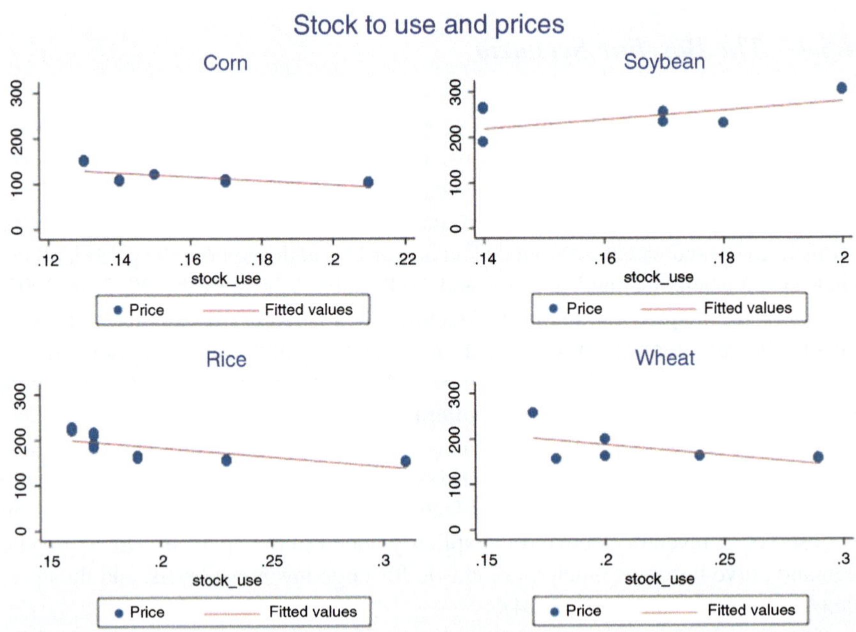

Fig. 4.4 Crop prices and the stock-to-use ratio

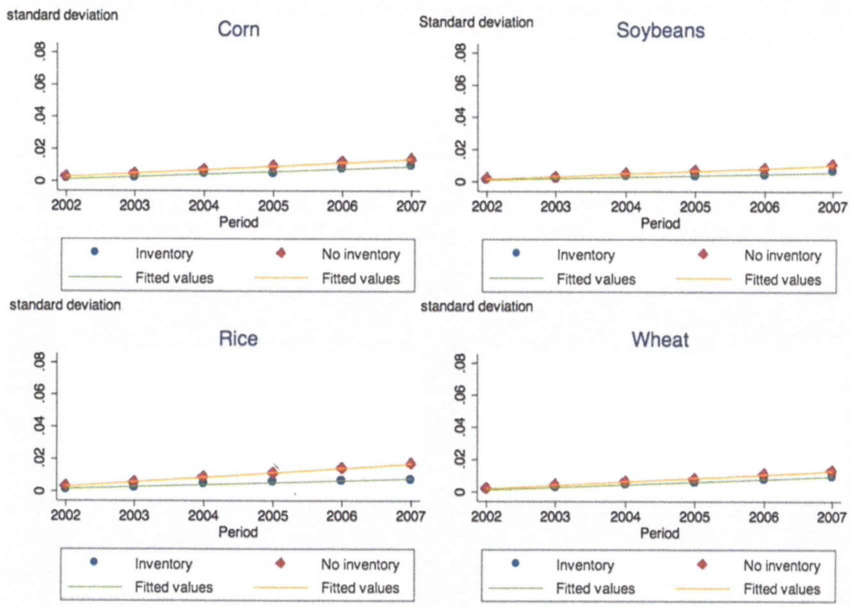

Fig. 4.5 The implication of demand and supply shocks on prices with and without inventory

price fluctuation in corn and soybeans is that consumption of corn for biofuel became significant around 2006, when the federal government began implementing biofuel mandates. Although biofuel subsidies have been in effect for several decades, mandates are the main cause for the recent increase in biofuel production. Furthermore, land allocated to corn replaces soybean land, resulting in higher soybean prices (not modeled explicitly, because we do not have data on land use). This complements the upward pressure on soybean prices attributed to biodiesel production. On the other hand, economic growth results in structural changes to demand in countries like China, where increased demand for feed led to larger demand for soybeans considerable growth of about 20 % between 2000 and 2008 was also observed for pork (Trostle 2008a, b).

Since it was assumed that rice and wheat are not utilized for biofuels in any significant quantities, and since land growing rice and wheat do not generally compete with corn, sugarcane, and oilseeds, the data reflect that the prices of rice and wheat are not influenced by biofuels. However, a general equilibrium framework, in contrast to the multi-market framework presented here, may identify indirect linkages between biofuel production and rice and wheat (Mitchell 2008). When the market for storage is excluded, higher price fluctuations are documented (Fig. 4.5 and Table 4.3). Graphing the standard deviation of prices for five crops for a represented shock, in Fig. 4.5, the shock caused prices to fluctuate more when inventory is not

Table 4.3 Contribution of various factors on increased price of selected food commodities (% price increase from counterfactual scenario in a given year)

	With inventory			Without inventory		
	Year			Year		
Crop	2005 (%)	2006 (%)	2007 (%)	2005 (%)	2006 (%)	2007 (%)
Biofuel shock						
Corn	4.40	6.80	9.80	5.50	7.40	9.80
Soybean	1.00	1.80	3.40	1.50	2.60	4.10
Rice	0.00	0.00	0.00	0.00	0.00	0.00
Wheat	0.00	0.00	0.00	0.00	0.00	0.00
Income shock						
Corn	7.90	12.20	15.30	12.40	16.70	19.50
Soybean	6.30	8.90	14.70	12.10	15.60	22.10
Rice	11.60	13.50	16.10	20.90	27.90	35.10
Wheat	11.10	16.00	21.20	15.10	21.40	27.70
Exchange rate shock						
Corn	3.50	5.00	7.60	4.60	6.20	9.40
Soybean	1.00	2.40	5.30	1.40	3.80	7.90
Rice	3.30	4.00	6.50	6.70	8.30	14.40
Wheat	6.60	7.30	11.00	8.10	8.90	13.10
Energy price shock						
Corn	2.20	2.90	2.90	3.30	3.60	3.60
Soybean	1.90	2.40	2.40	3.60	4.00	4.00
Rice	3.00	3.00	3.00	2.40	2.60	2.60
Wheat	2.80	3.10	3.10	3.60	4.00	4.00
Aggregate effect of all four shocks						
Corn	18	27	36	26	34	42
Soybean	10	15	26	19	26	38
Rice	18	20	26	30	39	52
Wheat	20	26	35	27	34	45

modeled explicitly. This picture emerges for all shocks. Inventory specification matters. Introducing inventory demand alters outcomes. Now, since inventories are observed and found significant by the numerical model, the following sections present exogenous shocks and simulate the impacts upon inventory demand.

4.6 Sensitivity Analysis

Because the empirical estimation of the demand and supply parameters as well as the demand for inventory are challenging but are key steps to accurately measuring the factors causing the food inflation of 2007–2008, two additional scenarios were numerically simulated to further check the robustness of this study's conclusion.

Table 4.4 Comparison of results between main and sensitivity analysis (% change as compared to the counterfactual scenario in 2007)

Shock	Crop	Main analysis (%)	Sensitivity analysis (%)
Biofuel	Corn	9.8	12.7
	Soybean	3.4	3.7
Income growth	Corn	15.3	20.3
	Soybean	14.7	16.0
	Rice	16.1	17.2
	Wheat	21.1	25.8

4.6.1 Inelastic Scenario

Key parameters in our analysis and in simulation-based models in general are the elasticities, which are used to calibrate the demand and supply curves. The alternative specification, denoted the *inelastic scenario*, assumes lower elasticities. The elasticities used in the baseline scenario were obtained from well-known and widely used sources such as the FAPRI elasticity database and the USDA elasticity database.[1] However, according to several other researchers, the elasticities of supply and demand for agriculture are more inelastic than those reported in the above databases (Gardner (1987). In order that the elasticities are on average lower than those in the baseline scenario and also conservative, we chose own-price supply elasticities in the range 0.2–0.3 and own- price demand elasticities in the range −0.3 to −0.2. Employing these elasticities, we find that the main qualitative conclusions regarding the importance of the different shocks from the baseline scenario hold.[2]

Comparing the baseline scenario to the inelastic scenario results in the price changes summarized in Table 4.4. This comparison emphasizes the importance of obtaining good elasticity estimates. The more inelastic scenario results in a larger impact.

Finally, we simulate the model using the inventory demand parameters estimated in Carter et al. (2008). Results confirm the conclusions derived for the baseline scenario. The price effect now is marginally smaller for all shocks. This is because the estimates of the parameters of the inventory demand employed in the elastic scenario imply an inventory demand function that is on average more elastic compared to that suggested by parameters estimated by Carter et al. (2008).

---

[1] http://www.ers.usda.gov/Data/InternationalFoodDemand/.

[2] To this end, using world data on four major crops, namely, corn, soybeans, wheat, and rice from 1960 to 2007, Roberts and Schlenker (2010) estimate that short-term, own-price elasticity of supply and demand for calories from these crops is less than 0.15 and greater than −0.1, respectively.

4.7 Conclusions

This chapter has focused on four major factors widely agreed to be responsible for food commodity price increases—economic growth, biofuel expansion, exchange rate fluctuations, and energy price change. The study also captures the effect of inventory adjustments. Incorporating an empirically estimated inventory demand function into the market-clearing condition shows that the impact of inventory on prices increases as the level of inventory diminishes. In the absence of shocks attributable to the four factors mentioned above, in 2007 the prices of corn, soybean, rapeseed, rice, and wheat would have been 26–36 % lower than the observed prices in that year. On the other hand, if inventory demand was to be ignored, in 2007 the prices would have been 38–52 % lower than the observed prices in that year. Abstracting from considerations of inventory responses leads to predictions of larger price changes.

Because key parameters in this analysis included the elasticities used to calibrate the demand and supply curves, the authors performed several sensitivity analyses on these values. In these alternative scenarios, many inelastic curves were introduced and compared against a demand curve based on the inventory parameters from Carter et al. (2008). It is concluded that although the percentage changes vary between scenarios, the main conclusion is that the inventory matters do not change. The relative magnitude of the various shocks also does not change.

From a policy standpoint, the food crisis emphasizes both the importance of a proactive inventory management policy and the need for mechanisms. Policies need to either compensate the poor when prices rise to abnormally high levels or more directly mitigate spikes in food prices. Such mechanisms may include biofuel mandates that adjust automatically to the situation in food markets, as well as inventory management policies. Expanding agricultural supply through investment in research and development and introducing policies that would allow more effective utilization of existing technologies. Meanwhile, investing in outreach and infrastructure that will enhance productivity also reduces the likelihood of a food price spike.

Various limitations require note for this study. Firstly, some important crop-specific factors, such as weather and productivity shocks (especially for wheat) and trade policies (especially for rice), are not considered. Secondly, speculative trade is a complexity that was not included in this model. Thirdly, no cross-price elasticities were introduced, which may have caused this study to underestimate the impact of the different factors on prices (Hochman et al. 2012). Another potential limitation stems from the fact that each crop market was evaluated separately, rather than in an integrated grains trade.

Although the conclusion is robust to a broad range of assumptions about the price elasticity of supply and demand for crops and parameters of the inventory demand function, an important area of future work is the empirical estimation of these parameters. Identifying correctly the inventory demand curve is a challenge, and is a key step to accurately measuring the factors causing the food inflation of 2007–2008. In future work we plan to further investigate these relationships, and to introduce cross-price elasticities. Moreover, the study does not analyze the 2008–2012 period, which was characterized by strong commodity price volatility. Thus, a further study is imperative to generate more policy insights by extending this study, incorporating the factors excluded here and also covering the 2008–2012 period.

References

Abbott, P. C., Hurt, C. A., & Tyner, W. E. (2008). *What's driving food prices?* Oak Brook, IL: Farm Foundation.

Alston, J. M., Beddow, J. M., & Pardey, P. G. (2009). Agricultural research, productivity, and food prices in the long run. *Science, 325*(5945), 1209.

Arrow, K. J., & Debreu, G. (1954). Existence of an equilibrium for a competitive economy. *Econometrica: Journal of the Econometric Society, 22*(3), 265–290.

Baka, J., & Roland-Holst, D. (2009). Food or fuel? What European farmers can contribute to Europe's transport energy requirements and the Doha Round. *Energy Policy, 37*(7), 2505–2513.

Carter, C., Rausser, G., & Smith, A. (2008). Causes of the food price boom. *Agricultural and Resource Economics Update, 12*(2), 7.

Collins, K. (2008, June 19). *The role of biofuels and other factors in increasing farm and food prices: A review of recent developments with a focus on feed grain markets and market prospects.* Written as supporting material for a review conducted by Kraft Foods Global, Inc. of the current situation in farm and food markets.

De Hoyos, R. E., & Medvedev, D. (2009). Poverty effects of higher food prices—a global perspective. *World Bank Policy Research Working Paper*(4887).

FAO. (2008, July). Crop prospects and food situation. *Food and Agriculture Organization*, (3)

FAO. (2008). The State of Food and Agriculture—Biofuels: Prospects, risks and opportunities. *Food and Agricultural Organization.*

Fischer, G., Hizsnyik, E. Prieler, S., Shah, M., & van Velthuizen, H. (2009, May) *Biofuels and food security.* Prepared by the International Institute for Applied Systems Analysis (IIASA) for OPEC Fund for International Development (OFID).

Gardner, B. L. (1987). *The economics of agricultural policies.* New York: Macmillan.

Glauber, J. (2008). *Testimony of USDA Chief Economist Joseph Glauber before the Joint Economic committee of Congress on May 1.*

Graff, G., Hochman, G., & Zilberman, D. (2009). The political economy of agricultural biotechnology policies. *AgBioForum, 12*(1), 34–46.

Graff, G., Zilberman, D., & Bennett, A. B. (2009). The contraction of product quality innovation in agricultural biotechnology. *Nature Biotech, 27*(8), 702–704.

Hochman, G., Kaplan, S., Rajagopal, D., & Zilberman, D. (2012). Biofuel and food-commodity prices. *Agriculture, 2*(3), 272–281.

Hochman, G., Rajagopal, D., Timilsina, G., & Zilberman, D. (2011) *Quantifying the causes of the global food commodity price crisis.* World Bank Policy Research Working Paper.

Hochman, G., Rajagopal, D., & Zilberman, D. (2010a). The effect of biofuels on crude oil markets. *Agbioforum, 13*(2), 112–118.

Hochman, G., Rajagopal, D., & Zilberman, D. (2010b). Are biofuels the culprit: OPEC, food, and fuel. *American Economic Review: Papers and Proceedings, 100*(2), 183–187.

Hochman, G., Sexton, S., & Zilberman, D. (2010c). *The economics of trade, biofuel, and the environment* (CUDARE Working Papers Series 1100). Berkeley, CA: Department of Agricultural and Resource Economics and Policy, University of California at Berkeley.

IMF. (2008). *Food and fuel prices: Recent developments, macroeconomic impact, and policy responses*. Report prepared by the Fiscal Affairs, Policy Development and Review, and Research Departments.
Ivanic, M., & Martin, W. (2010). *Promoting global agricultural growth and poverty reduction.* The World Bank, 18 March 2010, Manuscript. Available at https://www.gtap.agecon.purdue.edu/resources/download/4883.pdf
Johnson, S. (2008). Commodity prices: outlook & risks. International Monetary Fund, Washington DC (April 25, 2008).
Lampe, M. (2006). *Agricultural market impacts of future growth in the production of biofuels.* Paris, France: Working party on Agricultural Policies and Markets.
Lazear, E. P. (2008, May 14). Testimony of Edward P. Lazear Chairman, Council of Economic Advisers Before the Senate Foreign Relations Committee Hearing on "Responding to the Global Food Crisis."
Lipsky, J. (2008). Commodity prices and global inflation. Council on Foreign Relations, New York.
Martinot, E., & Sawin, J. (2009). *Renewables global status report: 2009 update*. Washington, DC: REN21 Renewable Energy Policy Network/Worldwatch Institute.
Mitchell, D. (2008). *A note on rising food prices* (Policy Research Working Paper #4682, Development Prospects Group). Washington, DC: World Bank.
Organization for Economic Co-operation and Development/Food and Agriculture Organization of the United Nations (OECD-FAO). (2008). *OECD/FAO agricultural outlook 2008/2017.* Author: Rome, Italy.
Peters, M., Langley, S., & Westcott, P. (2009). Agricultural commodity price spikes in the 1970s and 1990s: Valuable lessons for today. *Amber Waves, 7*, 16–23.
Rajagopal, D., Sexton, S., Hochman, G., Roland-Holst, D., & Zilberman, D. (2009). Biofuel and the food versus fuel trade-off. *California Agriculture, 63*(4), 20.
Rosegrant. M. W. (2008). *Biofuels and grain prices: impacts and policy response*. Testimony for the US Senate Committee on Homeland Security and Governmental Affairs. Washington, DC: IFPRI.
Schnepf, R. (2004). *Energy use in agriculture: Background and issues*. Report of the US Congressional Research Service and The Library of Congress.
Sexton, S., Zilberman, D., Rajagopal, D., & Hochman, G. (2009). The role of biotechnology in a sustainable biofuel future. *AgBioForum, 12*(1), 1–11.
The World Bank. (2008). *World development report 2008: Agriculture for development*. Author: Washington, DC.
Timilsina, G. R., & Shrestha, A. (2011). How much hope should we have for biofuels? *Energy, 36*, 2055–2069.
Timmer. C. P. (2008). *Causes of high food prices* (Asian Development Bank Economics Working Paper, 128).
Trostle, R. (2008a). Fluctuating food commodity prices a complex issue with no easy answers. *Amber Waves, 6*(5), 10–17.
Trostle, R. (2008b) *Global agricultural supply and demand: factors contributing to the recent increase in food commodity prices* (Outlook Report No. WRS-0801). Economic Research Service, US Department of Agriculture.
Williams, J. C., & Wright, B. D. (1991). *Storage and commodity markets*. Cambridge, MA: Cambridge University Press.
Zilberman, D., Hochman, G., Rajagopal, D., Sexton, S., & Timilsina, G. (2012). The impact of biofuels on commodity food prices: assessment and findings. *American Journal of Agriculture Economics, 95*(2), 275–281.
Zilberman, D., Khanna, M., & Scheffran, J. (Eds.). (2010). *Handbook of bioenergy economics and policy*. New York: Springer Science+Business Media.

Chapter 5
Economic Impacts of Biofuels

Govinda R. Timilsina

5.1 Introduction

The impacts of biofuels are not limited to biofuel industry and the agriculture sector, they spill over throughout an economy due to the inter-linkages between production sectors. The impacts are also felt across the borders through the international trade. For example, EU's mandate on biofuels could boost Brazil and Indonesia's economy. Using global macroeconomic models, particularly, computable general equilibrium (CGE) models, a number of studies have assessed economic impacts of EU's or global mandates on biofuels (e.g., Timilsina et al. 2012; Hertel et al. 2010; Kretschmer et al. 2010). While a single biofuel project implemented in a country may not have economic impacts at a scale noticeable at a national and an international level, a large group of projects or a biofuel policy aiming a large-scale expansion of biofuels would certainly have significant economic impacts at country level if not at global level. For example, Timilsina et al. (2012) finds that if the biofuel mandates and targets announced by the 40 plus countries around the world are executed by 2020 thereby increasing the share of biofuels in the global liquid fuel demand for transportation to 9 % from the current level of 3 %, various countries or regions would exhibit significant difference in their economic impacts ranging from 0.23 % loss of GDP in India to 0.05 % increase in GDP in Thailand.[1] However, at the global level the impact was fairly modest (0.02 %) compared to that in the baseline. Similarly, Kretschmer et al. (2009) find a 10 % EU-wide biofuel mandate not

[1] The change in GDP was measured compared to the baseline where the share of biofuels in total liquid fuel consumption for transportation was 5.4 %.

G.R. Timilsina (✉)
The World Bank, 1818 H Street, N.W., Washington, DC 20433, USA
e-mail: gtimilsina@worldbank.org

G.R. Timilsina and D. Zilberman (eds.), *The Impacts of Biofuels on the Economy, Environment, and Poverty*, Natural Resource Management and Policy 41,
DOI 10.1007/978-1-4939-0518-8_5,

causing a noticeable change in aggregate welfare of EU countries compared to a reference scenario where EU meets its 20 % GHG mitigation targets without biofuel mandate. However, the welfare impacts differ significantly across EU countries with some countries gaining, while others loosing.

This chapter aims to elaborate the long-term economic impacts of biofuels estimated by using CGE models based on existing literature, particularly Timilsina et al. (2012), which measures the impacts of a scenario that considers the implementation of biofuels mandates and targets announced by both developed and developing countries by the year 2020. The impacts are measured by comparing the scenario with the baseline which assumes that biofuels policies implemented prior to year 2009 will continue throughout the study period (2010–2020). Although a number of existing literature (e.g., Padella et al. 2012; Hertel et al. 2010; Taheripour et al. 2010; Banse et al. 2008) have analyzed economic impacts of biofuels, they have not reported detailed results; they are more limited on the impacts on agriculture commodities. This chapter goes beyond the agriculture sector and illustrates impacts on the entire economy of various countries and regions.

5.2 Methodology

Most studies assessing economic impacts of biofuels use CGE models. While the basic principle of CGE models is the same, they could differ on several aspects. For example, different models use different year's data, mainly the social accounting matrix (SAM), to calibrate their key parameters. The functional forms used to represent behaviors of production sectors are different across models. The same is true to represent household behavior. The values of elasticity of substitution could be different across the models. The scenarios simulated in the models are different. Nevertheless, most models report economic impacts of policies to promote biofuels (e.g., biofuel mandates) in terms of changes in GDP or economic welfare compared to that in a situation in the absence of such policies (i.e., baselines). Box 5.1 presents a brief summary of the CGE model and biofuel expansion scenarios simulated in Timilsina et al. 2012. Detailed descriptions of the model and data are available in Timilsina et al. (2010, 2012).

This model computes which and where biofuel feedstocks grow most efficiently, so as to reallocate land for optimal returns. For those countries (Brazil, the US, Malaysia, and South Africa) that have already or will fulfill energy targets prior to 2020, there will be no change in their targets until 2020. Countries are assumed to implement their mandates and targets linearly between 2009 and 2020 unless implementation schedules are defined otherwise. For example, some countries introduce a 5 % mandate in a given year, say 2015. In such a case, the model follows the actual implementation schedule.

Box 5.1 A brief description of CGE model used in Timilsina et al. (2012)

- The model is a multi-sector (28 sectors/commodities), multi-region (25 countries or regions), global recursive dynamic CGE model
- A nested constant elasticity of substitution (CES) functional form is used to represent production behavior in each sector; all biofuel feedstocks, biofuels, and energy commodities are separately represented
- The model represents land allocation through constant elasticity of transformation (CET) functional forms; each type of lands (crop, pasture, forest) are supplied from 18 agro-ecological zones in each countries/region
- A representative household maximizes its utility, using a non-homothetic constant difference of elasticity (CDE) functional form
- Imports are modeled by a system of Armington demands (CES functional forms); exports are depicted by CET functions; bilateral trades are also captured
- Representation of capital stock by vintage where new capital corresponds to the capital investments at the beginning of the period and old corresponds to the capital installed in previous periods; investment in each period is financed through household savings, government savings, and foreign borrowings
- Population and productivity growth are exogenous drivers for model dynamics
- GTAP database (version 7) with social accounting matrices (SAMs) for year 2004 was used

Although a country introduces a mandate or target for biofuel blending, an economic instrument needs to be provided to facilitate the realization of the mandate or target. Timilsina et al. (2012) assumes that a subsidy to biofuels is provided; the subsidy is financed by taxing gasoline and diesel, the fossil fuel counterparts of biofuels. Table 5.1 presents biofuel mandates and targets, subsidy required to realize the mandates and targets and fossil fuel tax rate required to finance the subsidy. The announced biofuel targets comprise 9 % of global liquid fuel consumption by the transportation sector in 2020. Even in the baseline, the penetration of biofuels would increase to 5.4 % by 2020. This is the most common approach in the literature to represent biofuels mandates and targets in a CGE model despite the fact that a mandate and subsidy impacts an economy differently. A mandate would increase the price of biofuel blends assuming that biofuels cost more per unit of energy delivery compared to their petroleum counterparts. Thus a mandate directly passes the cost burden to consumers. On the other hand, a government subsidy passes to the consumers indirectly if the government introduces taxes to finance subsidies.

Table 5.1 Biofuel shares, required subsidies and taxes on gasoline and diesel required to finance the subsidies in the year 2020

	Biofuel share (%)[a]		Subsidy rate (%)	Tax rate (%)
Country/Regions	Baseline	Scenario	Scenario	Scenario
Australia and New Zealand	0.56	1.23	36.71	0.16
Japan	0.48	0.60	14.85	0.05
Canada	1.47	4.10	50.07	0.56
United States[b]	3.91	4.07	0.96	0.04
France	4.54	10.00	58.36	1.11
Germany	5.86	10.00	43.35	1.03
Italy	2.54	10.00	65.32	1.21
Spain	2.31	10.00	61.01	1.12
UK	0.98	10.00	74.00	1.00
Rest of EU and EFTA	1.45	10.00	75.62	1.09
China	2.36	3.65	22.57	0.45
Indonesia	3.34	5.00	22.10	3.39
Malaysia	1.75	1.81	1.79	0.02
Thailand	1.86	5.20	51.92	0.92
Rest of East Asia and Pacific	0.60	1.49	42.45	0.22
India	4.86	16.70	53.26	3.35
Rest of South Asia		–	–	–
Argentina	1.61	5.00	52.27	0.87
Brazil[c]	18.77	18.77	–	–
Rest of LAC	1.32	1.48	16.74	0.10
Russia		–	–	–
Rest of ECA		–	–	–
MENA		–	–	–
South Africa	1.91	2.00	0.88	0.03
Rest of Sub-Saharan Africa		–	–	–

Source: Timilsina et al. (2012)
Note: The biofuels' share refer to the fraction of total liquid fuel demand for transportation in 2020 that will be met by biofuels when the targets and mandates are executed
EFTA European Free Trade Association, *LAC* Latin America and Caribbean, *EAC* Eastern Europe and Central Asia, *MENA* Middle East and North Africa
[a]Refers to the ratio of consumption of ethanol and biodiesel to consumption of ethanol, biodiesel, gasoline, and diesel in road transportation (the consumption is expressed in energy unit)
[b]Mandates for cellulosic ethanol are not included
[c]Since the biofuels' penetration in the scenario is equal to that in the baseline, no additional subsidy is required

5.3 Impacts on GDP

Gross domestic product is one of the main indicators to measure economic impacts of a policy shock (here biofuel mandates and targets). However, impacts on GDP of the global mandates and targets of biofuel are rarely available as a very few existing studies report these impacts. Several studies focus their impacts on agriculture sectors only (e.g., Hertel et al. 2010; Taheripour et al. 2010; Banse et al. 2008; Gohin 2008).

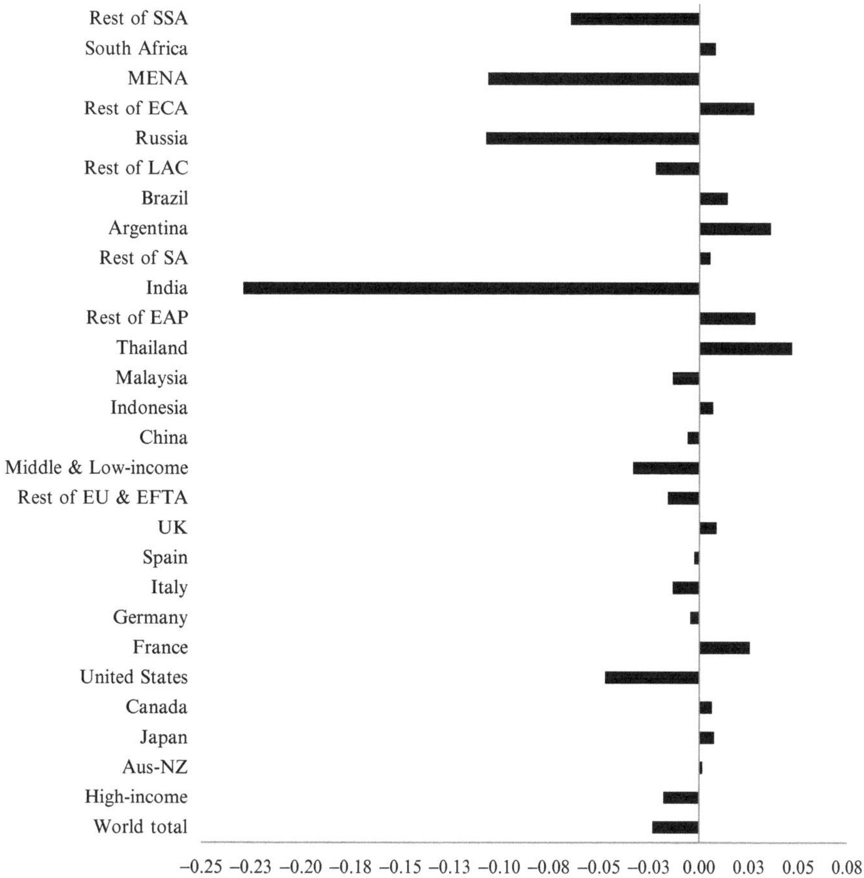

Fig. 5.1 Impacts on GDP of meeting global biofuels mandates and targets in 2020 (% change from the baseline). *Source*: Timilsina et al. (2012)

Figure 5.1 presents an example of GDP impacts of global biofuel mandates and targets. As illustrated in the figure, an expansion of biofuels to meet global biofuel mandates and targets would have relatively modest impacts on global GDP, about a drop of 0.03 %. However the impacts vary substantially across countries and regions depending on target levels and flexibility in meeting the target. India has an ambitious target of meeting 20 % (or 17 % in terms of energy content) of its liquid fuel demand for transportation through biofuels by 2020. Realizing this target would mean huge increase in imports of agricultural commodities (by about 4 %) and decrease in its agriculture commodity exports by about 2 %, thereby deteriorating terms of trade. On the other hand, countries where biofuel industry has already matured, such as Argentina, Brazil, Indonesia, and Thailand could experience an increase in GDP. Note however that the magnitudes of GDP increase in these countries are very small (less than 0.05 %).

5.4 Impacts on Economic Welfare

Economic welfare can be considered a better indicator than GDP to represent economic impacts. This is because GDP does not account for values of non-marked commodities or services, such as value of time spent on leisure. Hicksian equivalent variation[2] is normally used in CGE models to measure welfare impacts of any policy shock. It is defined as the value of a change in consumers' utility in monetary terms due to a policy shock. Hertel et al. (2010) finds that increasing US and EU's biofuel shares in the total liquid fuel consumption to 5.75 % and 6.25 %, respectively by 2015 would cause US$43 billion welfare loss globally. Most of the loss occurs in EU and oil exporting countries. Brazil and some other countries would experience some welfare gain. For the same policy, however, Padella et al. (2012) finds the global welfare loss more than double, US$96.6 billion. The second study uses a different closing rule in their CGE model than what the first study uses.

5.5 Impacts on Sectoral Outputs

Sectoral outputs are other common indicators to measure economic impacts of policy shocks in CGE modeling. Several studies (e.g., Hertel et al. 2010; Taheripour et al. 2010; Huang et al. 2012) report impacts of biofuel policies on sectoral outputs, particularly agricultural outputs, as the agriculture sector experiences the highest impacts due to expansion of biofuels. Table 5.2 illustrates the effects of biofuel mandates and targets on sectoral outputs at the aggregated (i.e., global, high income and low and middle income) level. Table 5.3 provides more insights on the effects on the biofuel sector and agriculture sub-sectors by countries/region. The output impacts show five distinct patterns: biofuels, agriculture commodities that are used as biofuel feedstock, petroleum products, and the rest of the commodities and services. As expected, the global biofuel mandates and targets cause a huge increase of biofuel production (64.5 % in 2020 compared to the baseline for the same year). Production of agriculture commodities that are used as biofuel feedstock increase significantly, for example, production of sugar crops increases by 8 %. Production of gasoline and diesel decreases thereby decreasing production of crude oil. Production of manufacturing, mining sectors and service sectors also drops down though slightly. One exception is chemical industry; it mostly uses petrochemicals, which are not displaced by biofuels.

Due to the comparative advantages and policy targets, Brazil, China, France, and India would realize relatively greater biofuel production than other countries (Table 5.3). The percentage change numbers in Table 5.3 may not reflect this. Brazil

[2] The concept of equivalent variation was introduced by British economist Sir John Richard Hicks in 1937. For more details, please refer to Mas-Colell et al. (1995).

Table 5.2 Change in sectoral outputs in 2020 due to biofuel mandates and targets

	Global	High income	Low and medium income
Total biofuels	64.5	89.6	42.1
Ethanol	55.5	72.4	42.7
Biodiesel	143.6	165.1	22.0
Total agriculture	0.4	0.8	0.2
Paddy rice	–0.2	0.1	–0.2
Sugar crops	8.1	10.1	7.4
Fruits and vegetables	–0.1	0.0	–0.1
Wheat	1.3	2.6	0.4
Corn	1.0	0.7	1.3
Other coarse grains	4.5	11.1	0.3
Oilseeds	2.4	6.3	0.3
Livestock	0.0	0.0	–0.1
Total industry and service	–0.05	–0.05	–0.04
Processed food	0.02	0.14	–0.19
Forestry	–0.09	–0.17	–0.04
Coal	–0.15	–0.07	–0.19
Crude oil	–0.34	–0.26	–0.35
Natural gas	0.02	–0.10	0.07
Other mining	–0.05	–0.14	–0.01
Gasoline	–1.07	–0.80	–1.32
Diesel	–0.61	–1.08	–0.31
Refined oil	0.12	–0.07	0.22
Chemicals	0.07	0.05	0.09
Other manufacturing	–0.07	–0.11	–0.03
Electricity	–0.07	–0.05	–0.09
Gas distribution	–0.11	–0.13	–0.08
Construction	–0.12	–0.18	–0.05
Transport services	0.00	0.05	–0.07
Other services	–0.03	–0.03	–0.02

is driven mostly by international trade demands, while other nations are primarily spurred by energy mandates and targets. Following biofuel expansion, each region experiences agricultural expansion. Trade demands also add force to the agricultural production (Table 5.3). However, rice and fruits and vegetables,[3] experience drop in their outputs as some lands used by these crops are now used for the production of biofuel feedstock. The EU countries show largest relative increases in agricultural output, shown in Table 5.3. Middle and low-income countries experience limited expansion in grains, since sugarcane and other competing crops remain more profitable in these regions. Corn production in China is excepted as it grows manifold under the scenario.

[3] Rice production decreases as land is reallocated towards oilseeds, sugar, and grains. Behaving as an inferior good, rice exhibits a negative price response to income.

Table 5.3 Change in biofuel and agricultural sectors' outputs (% change relative to the baseline in 2020)

	Biofuel	Agriculture total	Paddy rice	Sugar crops	Fruits and vegetables	Wheat	Corn	Other grains	Oil seeds	Livestock
Australia and New Zealand	20.9	0.3	0.1	0.4	0.3	1	1.2	0.9	1.9	0.2
Japan	3.8	0.3	0.1	0.5	0.4	0.9	0.4	1.1	0.9	0.3
Canada	45.0	0.3	0	0.2	0	0.7	4	1	1.1	0
United States	1.1	0.2	0.1	0.1	0.1	1	0.7	0.7	0.9	0.1
France	268.0	2.6	−2	68.8	−1.4	5.2	0.2	8.2	28.3	−0.5
Germany	112.0	1.3	0	0.7	0	7.1	1.8	7.7	17.5	0.1
Italy	321.0	1.1	0.6	0.4	0.4	2.2	0.8	2.1	14.1	0.4
Spain	367.5	1.5	−0.1	0.5	0.2	2.6	1.3	36.8	3.2	0.2
UK	500.1	1.1	0	19.9	−0.7	−1.4	0	62.1	11.2	−0.5
Rest of EU and EFTA	538.8	0.7	0.4	3	0.2	2.1	−0.5	9.3	7.9	0
China	55.4	0.2	−0.1	0.2	−0.1	1.1	4.7	−0.2	−0.2	−0.1
Indonesia	49.2	0.1	−0.1	8.1	−0.2	0	0	0	0.1	−0.1
Malaysia	0.5	0.1	0.2	0.2	0.1	0	0	1.4	0.4	0
Thailand	88.1	0.4	−0.9	28	−0.7	0	0.4	−0.3	−0.5	−0.5
Rest of East Asia and Pacific	53.7	0.1	0	3.2	0.1	0.4	0.9	0.6	0.3	0
India	205.8	0.2	−0.7	13.3	−1	−0.4	−0.4	−0.4	−0.6	−0.6
Rest of South Asia	58.1	0.1	0	0.9	0	0.3	0.2	0.3	1	0
Argentina	45.9	0.4	0.6	−0.1	0.1	0.5	1.5	0.3	0.5	−0.2
Brazil	31.1	0.3	−1.2	17.9	−1.4	−1.2	−1.4	−0.9	1	−1
Rest of LAC	4.2	0.1	0	0.2	0.2	0.5	0.1	0.1	1.7	0
Russia	−1.3	0.2	0.1	0.3	0.4	0.3	0.4	0.6	1.7	0
Rest of ECA	−0.9	0.3	0.1	0.5	0.3	0.7	0.7	0.8	1.9	0.1
MENA	−1.2	0.3	−0.1	0.6	0.3	1.1	1.4	0.9	1.4	−0.1
South Africa	0.5	0.4	0	0.6	0.5	1.8	0.4	1.4	0.7	0.1
Rest of Sub-Saharan Africa	0.1	0.1	0	0.3	0.2	1	0	0	0.2	−0.2

Source: Timilsina et al. (2012)

5.6 Impacts on International Trade

The global biofuel mandates and targets cause a large expansion of international trade of biofuels. At the global level, the value of biofuel trade increases by more than 2.5 times from the baseline scenario in 2020 (Table 5.4). Some countries such as United Kingdom, Scandinavian countries, India, Spain, and Italy would face a large increase in their imports of biofuels to meet their demand. Germany, France, Spain, and United Kingdom would be the major countries experiencing increased exports of biofuels.

Compared to that biofuels, percentage change in international trade of other commodities is smaller for two reasons. First, the base of international trade of

Table 5.4 Impacts of global biofuels mandates and targets on aggregated international trade (% change from the baseline in 2020)

	Imports			Exports		
Country/Region	Biofuels	Agriculture	Others	Biofuels	Agriculture	Others
World total	258.7	1.0	–0.2	258.7	1.0	–0.2
High-income	310.9	2.3	–0.2	370.6	0.9	–0.1
Australia–New Zealand	n.a	0.8	0.1	n.a	1.0	–0.1
Japan	–1.6	0.5	–0.1	0.0	2.0	–0.1
Canada	65.5	1.3	0.0	0.3	1.0	–0.1
United States	0.6	0.6	–0.3	38.3	1.1	0.4
France	153.8	4.5	0.1	486.1	0.5	–0.8
Germany	78.9	5.0	–0.1	873.5	1.1	–0.1
Italy	319.7	1.7	–0.2	0.0	1.0	–0.3
Spain	362.1	2.8	–0.1	375.9	1.2	–0.4
UK	1042.4	2.8	–0.1	472.7	–1.1	–0.1
Rest of EU and EFTA	637.2	2.7	–0.2	82.7	0.6	–0.2
Middle and low-income	203.6	0.7	–0.2	181.1	1.4	–0.3
China	0.0	0.6	–0.1	25.9	1.0	0.0
Indonesia	0.0	1.1	–0.1	1.3	0.3	–0.1
Malaysia	0.0	0.9	–0.1	1.7	1.0	0.0
Thailand	0.0	1.0	–0.2	–39.0	–0.1	–0.2
Rest of EAP	54.5	0.7	0.0	0.0	1.4	0.0
India	420.3	3.9	–1.3	0.0	–6.0	–0.7
Rest of SA	0.0	0.4	–0.1	306.1	2.7	–0.2
Argentina	0.0	0.7	0.0	1.5	1.5	–0.3
Brazil	0.0	1.2	1.3	198.5	0.5	–3.1
Rest of LAC	0.0	0.6	–0.1	0.0	1.4	–0.2
Russia	0.0	–0.6	–0.5	0.0	3.7	–0.4
Rest of ECA	–5.8	0.0	0.0	0.0	1.7	–0.1
MENA	–2.2	–0.3	–0.7	0.0	2.2	–0.7
South Africa	0.0	0.0	0.0	0.0	1.4	–0.1
Rest of SSA	–3.6	–0.1	–0.3	0.0	2.2	–0.4

those commodities is already large so percentage change would not be large even if value of trade increases significantly. Second, unlike the case of biofuels where the mandates and targets have direct impacts, international trade of other commodities gets affected only indirectly. Countries like Thailand would reduce their exports of biofuels to meet their own mandate.

The changes in agriculture industry and trade indicate new global market dynamics. Tables 5.5 and 5.6 present change in imports and exports of agriculture commodities. All countries with biofuel targets experience greater trade of agricultural commodities. Relatively higher imports are demanded by the EU countries and India than by other nations. Imported crop types are relatively higher for sugar and oilseeds. On the export side, main biofuel feedstock providers include Brazil, Argentina, Russia, and developing regions of Sub-Saharan Africa, MENA, and South Asia (excluding India). Multilateral trading weighs upon variables of

Table 5.5 Impacts of global biofuels mandates and targets on imports of agriculture commodities (% change from the baseline in 2020)

Country/region	Paddy rice	Sugar crops	Fruits and vegetables	Wheat	Corn	Other grains	Oilseeds	Livestock
World total	0.3	1.6	0.7	0.9	1.8	4.3	4.2	0.4
High-income	0.4	14.0	0.8	3.0	1.6	14.0	13.3	0.5
Australia–New Zealand	0.8	−0.6	0.8	2.8	1.1	−0.3	1.5	0.5
Japan	0.2	−0.9	−0.2	0.5	0.8	0.4	0.8	−0.1
Canada	0.5	0.5	0.8	9.3	9.1	1.7	1.4	0.6
United States	0.3	−3.1	0.7	0.6	0.2	0.0	1.0	0.1
France	1.1	211.2	2.1	34.2	1.4	−1.3	84.6	1.2
Germany	0.5	0.0	0.7	14.6	1.4	7.9	32.3	0.4
Italy	0.1	−2.0	0.6	0.2	0.3	0.6	18.4	0.8
Spain	0.9	−0.2	0.5	−0.7	0.6	43.5	1.2	0.2
UK	0.4	19.1	1.2	4.5	1.2	133.6	20.5	1.2
Rest of EU and EFTA	0.3	4.5	0.6	4.5	2.3	15.5	13.2	0.4
Middle and low-income	0.3	1.4	0.7	0.5	1.8	−0.3	1.2	0.4
China	0.2	−0.1	0.5	1.7	14.2	1.7	1.1	0.4
Indonesia	1.5	67.5	0.9	0.8	0.6	0.7	2.0	0.8
Malaysia	1.7	1.1	0.6	1.1	1.0	0.9	1.4	0.3
Thailand	1.9	159.3	1.1	0.2	2.5	−0.5	1.0	0.6
Rest of EAP	0.4	15.4	0.7	0.8	0.9	0.2	0.9	0.3
India	6.8	173.6	4.3	6.2	6.4	5.1	6.1	2.1
Rest of SA	0.3	4.4	0.5	0.0	0.4	0.0	0.6	0.2
Argentina	−1.8	0.3	0.4	0.0	9.6	0.1	0.8	0.5
Brazil	1.9	0.0	1.8	0.3	1.9	1.3	3.8	1.2
Rest of LAC	0.5	0.5	0.6	0.4	0.3	0.3	1.3	0.3
Russia	−0.5	−2.1	−0.1	−1.4	−0.9	−3.6	−1.5	−1.1
Rest of ECA	0.1	−1.9	0.4	−1.5	−0.5	−3.8	0.6	0.2
MENA	0.2	−1.7	0.1	−1.3	0.1	−0.9	0.0	−0.2
South Africa	0.7	−0.7	0.3	−0.5	−0.8	−0.9	0.4	0.2
Rest of SSA	−0.2	−0.5	−0.2	0.6	−0.7	−2.8	−1.2	−0.5

Source: Timilsina et al. (2012)

commodity baskets and distance between producers and demanders. Lesser barriers and higher volume of trade can bring fluidity to large nations, where previously a domestic market may have had more trouble transferring surpluses to deficit areas.

5.7 Impacts on Commodity Prices

A body of literature focuses on the impacts of biofuels on food prices. However, biofuels could impact prices of other commodities such as fossil fuels, manufacturing goods and services. A CGE model can measure the change in prices of goods

Table 5.6 Impacts of global biofuels mandates and targets on exports of agriculture commodities (% change from the baseline in 2020)

Country/region	Paddy rice	Sugar crops	Fruits and vegetables	Wheat	Corn	Other grains	Oilseeds	Livestock
World total	0.35	1.57	0.74	0.93	1.77	4.33	4.17	0.41
High-income	0.65	−2.64	0.64	0.71	1.98	4.22	4.26	0.39
Australia–New Zealand	0.63	2.87	0.99	1.62	7.10	1.98	3.16	0.61
Japan	2.48	4.67	2.39	3.41	0.96	2.94	5.61	1.42
Canada	0.00	2.01	0.76	1.36	4.00	2.08	2.52	0.30
United States	0.65	6.93	0.70	2.05	2.13	1.41	2.47	0.78
France	−1.56	−44.07	−1.91	−1.31	1.42	13.86	13.03	−0.93
Germany	0.00	6.09	0.50	0.30	4.27	10.87	4.07	0.55
Italy	1.13	4.59	1.16	4.13	2.70	5.63	9.68	0.21
Spain	0.31	39.68	0.93	6.01	5.81	7.20	13.76	0.77
UK	0.00	−28.79	−0.91	−1.32	0.00	−25.07	11.62	−1.02
Rest of EU and EFTA	1.02	−1.29	0.85	−1.36	−1.46	1.31	6.17	0.25
Middle and low income	−0.23	4.35	0.90	2.45	1.17	5.51	4.07	0.67
China	0.62	8.23	1.28	0.55	−12.05	−0.05	3.08	0.57
Indonesia	0.32	−52.04	0.41	0.00	9.19	0.00	2.66	−0.32
Malaysia	−0.62	1.21	1.16	0.00	0.00	2.11	3.04	0.64
Thailand	−1.94	−69.79	−0.11	0.00	2.99	−0.07	0.47	−0.21
Rest of EAP	1.05	−16.39	1.30	2.00	9.88	4.17	3.06	0.73
India	−5.69	−79.74	−6.25	−5.92	−6.92	−8.83	−2.35	−2.85
Rest of SA	1.45	11.89	2.66	3.82	6.27	8.92	3.56	1.39
Argentina	4.21	0.79	1.05	1.37	−0.06	3.91	2.50	0.13
Brazil	−2.02	−49.98	−1.90	1.12	−1.08	2.77	4.12	−1.97
Rest of LAC	0.81	2.80	1.07	3.35	5.34	5.56	5.93	0.63
Russia	2.30	0.00	2.87	7.47	8.48	8.39	13.05	2.92
Rest of ECA	1.20	5.38	1.34	2.66	2.80	5.60	7.44	0.83
MENA	1.16	4.69	1.88	5.16	7.53	7.26	9.03	1.67
South Africa	0.00	4.56	1.43	3.11	0.92	3.11	4.18	0.73
Rest of SSA	1.71	6.09	2.08	3.48	5.79	7.27	4.87	1.73

Source: Timilsina et al. (2012)

and services. However, most studies analyzing impacts of biofuels do not report price impacts especially commodities other than agriculture. Sugar crops (both cane and beet) are the feedstock for ethanol, the increased demand for ethanol due to the mandates and targets put upward pressure on prices of sugar crops. Their price could be more than 7 % higher than that in the baseline in 2020. The mandates and targets would also cause biofuel prices to rise. The same would be the case for all agricultural commodities. Prices of fossil fuels, particularly crude oil and petroleum products, would fall as biofuels reduce their demand through substitution (Fig. 5.2).

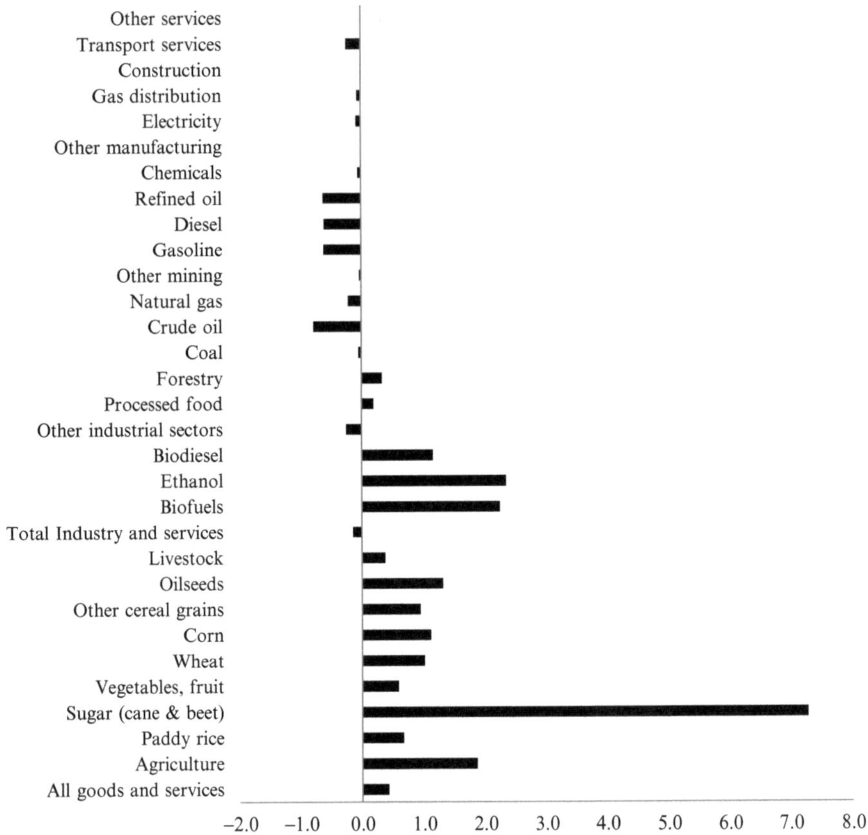

Fig. 5.2 Impacts of global biofuel mandates and targets on commodity prices (% change from the baseline in 2020)

5.8 Conclusions

This chapter attempts to illustrate the economic impacts of large-scale expansion of biofuels to meet the global mandates and targets by providing in-depth results from a CGE model used in Timilsina et al. (2012), and other relevant literature.

Under the scenario of meeting biofuel targets and mandates announced by 40 plus countries around the world, the expansion of biofuels could lead to significant impacts on overall economic outputs, sectoral outputs, international trade of goods and services, and prices of goods and services. The expansion of biofuels to meet the global mandates and targets does not affect the global economic output (GDP) noticeably, but the variations of impacts across countries and region are significant due to differences in the mandates and targets themselves and flexibilities these countries have to realize the mandates and targets. Countries like India with an ambitious target and low flexibility to domestically meet the increased demand for biofuel and feedstock experiences higher loss in its economic outputs compared to other countries. On the other hand, countries with already matured biofuel industries and no land constraints to increase their supply for exports, such as Brazil and Indonesia will see economic gains.

The global biofuel mandates and targets would heavily increase production or sectoral outputs from the biofuel industry and also increase production from the agricultural sector as a whole although sectoral outputs of non-biofuel feedstock (e.g., rice, fruits, and vegetables) drop. Biofuels replace petroleum products thereby causing them to lose their productions. It would also negatively impact the production of other goods and services.

Economic impacts are more driven by international trades than domestic supplies of biofuels and feedstock. International trades of biofuels (ethanol and biodiesel) would increase in many folds in several countries. International trade of agricultural commodities, especially biofuel feedstock, would also increase. However, international trade of other goods and services would drop. The global biofuel mandate and targets would cause an upward pressure to prices of goods and services except energy goods and services, whose prices would drop. Biofuels and major agricultural commodities, particularly those used as biofuel feedstock, would experience a significant increase in their prices. In overall, the economic impacts are mixed with some countries gaining while others losing. When aggregated, biofuels have a slight negative impact on the global economy.

References

Banse, M., van Meijl, H., Tabeau, A., & Woltjer, G. (2008). Will EU biofuel policies affect global agricultural markets? *European Review of Agricultural Economics, 35*(2), 117–141.

Gohin, A. (2008). Impacts of the European biofuel policy on the farm sector: A general equilibrium assessment. *Review of Agricultural Economics, 30*(4), 623–641.

Hertel, T. W., Tyner, W., & Birur, D. K. (2010). The global impacts of biofuel mandates. *The Energy Journal, 31*(1), 75–100.

Huang, J., Yang, J., Msangi, S., Rozelle, S., & Weersink, A. (2012). Biofuels and the poor: Global impact pathways of biofuels on agricultural markets. *Food Policy, 37*, 439–451.

Kretschmer, B., Peterson, S., & Ignaciuk, A. M. (2009). Integrating biofuels into the DART model: Analysing the effects of the EU 10 % biofuel target. In Loy, J.-P. Müller, & Rolf A. E. (Eds.), Agrar- und Ernährungsmärkte nach dem Boom: 49. Jahrestagung der Gesellschaft für Wirtschafts- und Sozialwissenschaften des Landbaues e. V. vom 30. September bis 02. Oktober 2009, ISBN 978-3-7843-5106-3, pp. 63–74.

Mas-Colell, A., Whinston, M., & Green, J. (1995). *Microeconomic theory*. New York: Oxford University Press.

Padella, M., Finco, A., & Tyner, W. E. (2012). Impacts of biofuels policies in the EU. *Economics of Energy & Environmental Policy, 1*(3), 87–104.

Taheripour, F., Hertel, T., Tyner, W., Beckman, J., & Birur, D. (2010). Biofuels and their by-products: global economic and environmental implications. *Biomass and Bioenergy, 33*(3), 278–289.

Timilsina, G. R., Beghin, J. C., van der Mensbrugghe, D., & Mevel, S. (2010). *The impacts of biofuel targets on land-use change and food supply: A global CGE assessment* (Policy research Working Paper 5513). Washington, DC: The World Bank.

Timilsina, G. R., Beghin, J. C., van der Mensbrugghe, D., & Mevel, S. (2012). The impacts of biofuel targets on land-use change and food supply: A global CGE assessment. *Agriculture Economics, 43*, 315–332.

Chapter 6
Biofuels and Poverty

Caesar B. Cororaton and Govinda R. Timilsina

6.1 Introduction

Several studies have looked at the aggregate economic effects of biofuels (IFPRI 2008; FAO 2008; Ivanic and Martin 2008; Mitchell 2008; Keyzer et al. 2008; Hochman et al. 2011; Timilsina et al. 2012), but only few research have focused on analyzing the income distribution and poverty effects of higher increased biofuel at the global level. deHoyos and Medvedev (2011) examine the poverty effects of higher biofuel production using a global CGE model, a model without an explicit representation of biofuel sectors and land-use. Runge and Senauer (2007) examine the impacts of biofuel promoting policies on food prices and poverty and find that policies that promote ethanol have adverse impact on food prices and poverty especially in developing countries. However, the results of some existing studies have reported opposite results. For example, using a computable general equilibrium (CGE) model for Mozambique, Arndt et al. (2008) find that higher sugar cane production generates favorable effects on growth and distribution; the effects on rent to land and wages are relatively higher if production is through smallholders than through large plantations. A study conducted for Mali using CGE and micro-simulation models, Boccanfuso et al. (2013) find that if the expansion of food crops for biofuel production does not compete for land use in agricultural production, agriculture as a whole improves slightly, but if it competes for land with other crops, then agriculture slightly declines. The distributional effects in Mali indicate that urban households who are net consumers of food loss with the expansion of biofuel production because of higher prices of staple food, but the rural households would gain.

C.B. Cororaton
Research Fellow, Global Issues Initiative/Institute for Society, Culture and Environment, Virginia Polytechnic Institute and State University, Virginia, USA
e-mail: ccaesar@vt.edu

G.R. Timilsina (✉)
The World Bank, 1818 H Street, N.W., Washington, DC 20433, USA
e-mail: gtimilsina@worldbank.org

G.R. Timilsina and D. Zilberman (eds.), *The Impacts of Biofuels on the Economy, Environment, and Poverty*, Natural Resource Management and Policy 41,
DOI 10.1007/978-1-4939-0518-8_6,

Hertel (2009) also finds that increased biofuel production results in higher factor returns in developing countries. Habib-Mintz (2010) finds that higher jatropha-based biofuel in Tanzania reduces poverty and food insecurity if strong regulatory frameworks for land, investment management, and rural development are in place.

Using global CGE and micro-simulation models, Cororaton and Timilsina (2012) simulated the global poverty and income distribution effects of the biofuel targets and mandates announced by 40 plus countries around the world. Based on that study, this chapter discusses the impacts of large-scale expansion on the poverty, labor migration, and income inequality.

6.2 Analytical Framework

Cororaton and Timilsina (2012) first use a global CGE model to analyze the impacts of biofuels. The model is the same as discussed in Chap. 5 earlier. The baselines and scenarios simulated are also the same as presented in Chap. 5. The detail description of the CGE model is available in Timilsina et al. (2012). For analyzing the impacts of biofuel expansion on poverty and income inequality, the key CGE results, such as commodity demand, commodity prices, factor prices, and household income, etc. are fed into another model, global income distribution model (GIDD).[1] GIDD uses household survey data of 116 countries, representing about 90 % of the world population. It projects household survey data using three sets of ex-ante macroeconomic information: (a) changes in demographic composition which consist of projection of population by age and by educational attainment; (b) movement of labor between agriculture and non-agriculture; and (c) economic growth.

The main sources of data in the GIDD model include: (a) the dataset assembled for the production of the World Bank World Development Report (WDR) for developing countries, which are drawn largely from the Living Standards and Measurement Study (LSMS) and the African Institute for Sustainability and Peace (ISP)-Poverty monitoring group; (b) the Europe and Central Asia (ECA) databank and the different World Bank sources for Eastern Europe countries, and (c) the Luxembourg Income Studies (LIS) database for most of the developed countries. The GIDD database covers all regions in the world. Eastern Europe and Central Asia is 100 % covered; Latin America 98 %; South Asia 98 %; East Asia and Pacific 96 %; High Income Countries 79 %; Sub-Saharan African 74 %; and Middle East and North Africa 70 % (Ackah et al. 2008).

Based on historical data, the GIDD model recalibrates the educational endowments of the population in some year in the future, which also changes the labor supply by age and skill groups in the CGE model. The CGE incorporates expansion of biofuel policy shocks and simulates the effects into the future on key economic variables such as real per capita GDP and per capita consumption, consumer price index of agriculture, and non-agriculture commodities, labor movement between

[1] For more discussion on the GIDD model, please refer to Ackah et al. 2008.

rural and urban and between agriculture and non-agriculture sectors, and changes in wages of various types of labor. These simulated economic effects are used in the GIDD model together with the new set of recalibrated weights. The GIDD model uses all this information to calculate the income distribution and poverty effects of large-scale expansion in biofuels.

The scenario simulated is the same as discussed in Chap. 5. While Chap. 5 presents macroeconomic effects, this chapter concentrates on distributional impacts.

6.3 Effects on World Prices of Feedstock and Food

Higher biofuel production increases the world prices of feedstock (Fig. 6.1)[2]. If the targets and mandates announced by 40 plus countries around the world are implemented, it would increase the demand for biofuel feedstock and their prices. For example, the price of sugar would increase by more than 7 % from the BaU scenario in 2020. There would also be noticeable increases in the prices of other feedstock such as oilseeds and corn.

Higher prices of feedstock lead to higher prices of food (Fig. 6.2). The increase in food CPI is higher in developing countries than in developed countries largely due to higher shares of food in the consumption basket of the consumers in the former than in the latter.

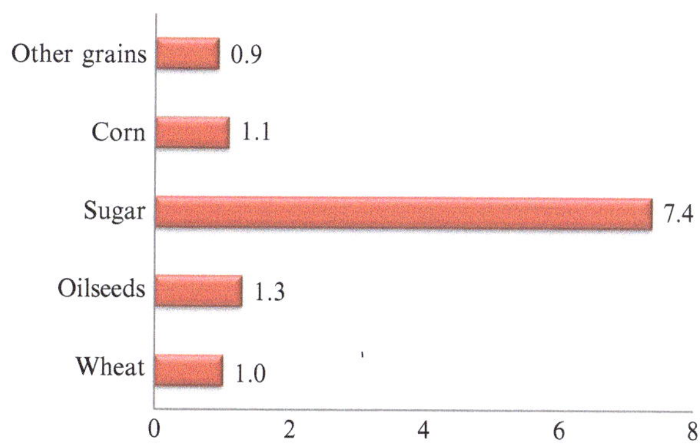

Fig. 6.1 World prices of feedstock due to biofuel mandates and targets in 2020 (% change from BaU)

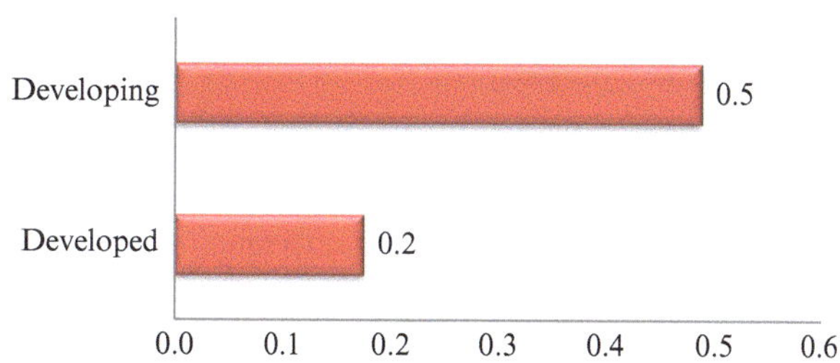

Fig. 6.2 Change in food consumer price index due to biofuel mandates and targets in 2020 (% change from BaU)

[2] *Source for Figs. 6.1–6.8*: Cororaton and Timilsina (2012.

6.4 Effects on Factor Prices and Income

However, higher biofuel production improves factor returns and income. Land rent income improves in all countries and regions as biofuel production increases (Table 6.1). The increase in land rent income is highest in France, followed by United Kingdom. Developing countries and regions also see higher land rent income from increased biofuel production. The impact on the total labor income is mixed. While several countries and regions see higher labor income as biofuel production increases, few countries register slight decline in labor income. However, the improvement in land rent income, more than, offsets the decline in labor in countries and region where labor income decreases.

The increase in consumer prices from higher biofuel production generates varied effects on real per GDP across countries and regions. While most factor returns and factor improve from higher biofuel production, some of these improvements are not high enough to offset the increase in consumer prices. Countries in Sub-Saharan Africa, Middle East and North Africa, Russia, India, and China show declining real per capita income. Most of the developed countries, however, have improving real per capita GDP (Fig. 6.3).

Table 6.1 Impacts on labor and land income due to biofuel targets and mandates in 2020 (% change from BaU)

	Land income	Labor income
	Scenario/BaU	Scenario/BaU
China	0.8	0.0
Japan	1.6	0.0
Indonesia	1.0	0.0
Malaysia	0.6	−0.1
Thailand	1.9	0.2
India	3.7	0.0
Canada	2.0	0.0
United States	1.4	−0.1
Argentina	1.8	0.1
Brazil	3.3	0.7
France	11.6	0.3
Germany	4.9	0.1
Italy	6.2	0.2
Spain	5.0	0.2
United Kingdom	9.6	0.1
Russia	0.8	−0.4
South Africa	1.2	0.1
Rest European Union and EFTA	2.9	0.0
Rest of Latin America and Caribbean	0.9	−0.1
Australia and New Zealand	1.2	0.1
Rest of East Asia and Pacific	0.7	0.1
Rest of South Asia	0.6	0.0
Rest of Europe and Central Asia	1.2	0.1
Middle East and North Africa	1.3	−0.7
Rest Sub-Saharan Africa	0.4	−0.3

Source: Cororaton and Timilsina (2012)

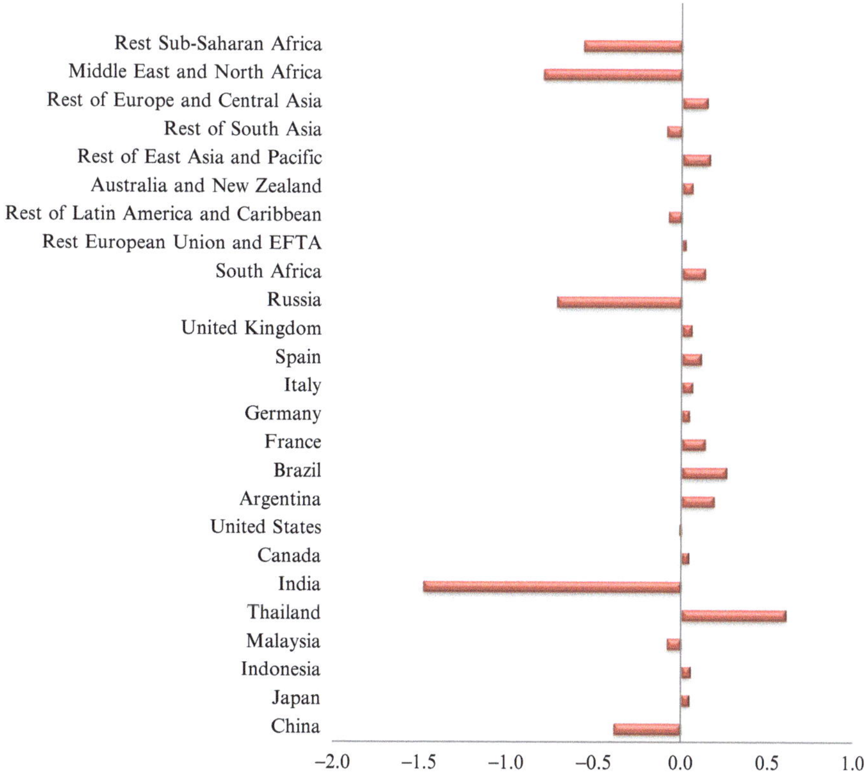

Fig. 6.3 Effects of biofuel mandates and targets on real per capita GDP in 2020 (% change from BaU)

6.5 Effects on Labor Movement

Feedstock production in developing countries is relatively unskilled labor intensive because of higher supply of unskilled labor rural areas. The expansion in biofuel production which increases the demand for feedstock increases the relative wages of unskilled labor, as well as the demand for unskilled labor in developing countries. Figure 6.4 shows the changes in the wage ratio between skilled and unskilled labor. One can observe that the wage ratio decreases, and the decline is relatively higher in developing countries compared to developed countries. This declining wage ratio implies that wages of unskilled labor are increasing faster than wages of skilled labor. The highest increase in the relative wages of unskilled rural labor is in India, Middle East and North Africa, Sub-Saharan Africa, Brazil, Argentina, Thailand, Indonesia, and the rest of Latin America.

The differential effects on wages affect labor movement of labor especially in developing countries. The expansion of biofuel production which leads to higher feedstock production decreases the demand for urban unskilled labor but increases the demand for rural unskilled in developing countries (Fig. 6.5). This implies labor

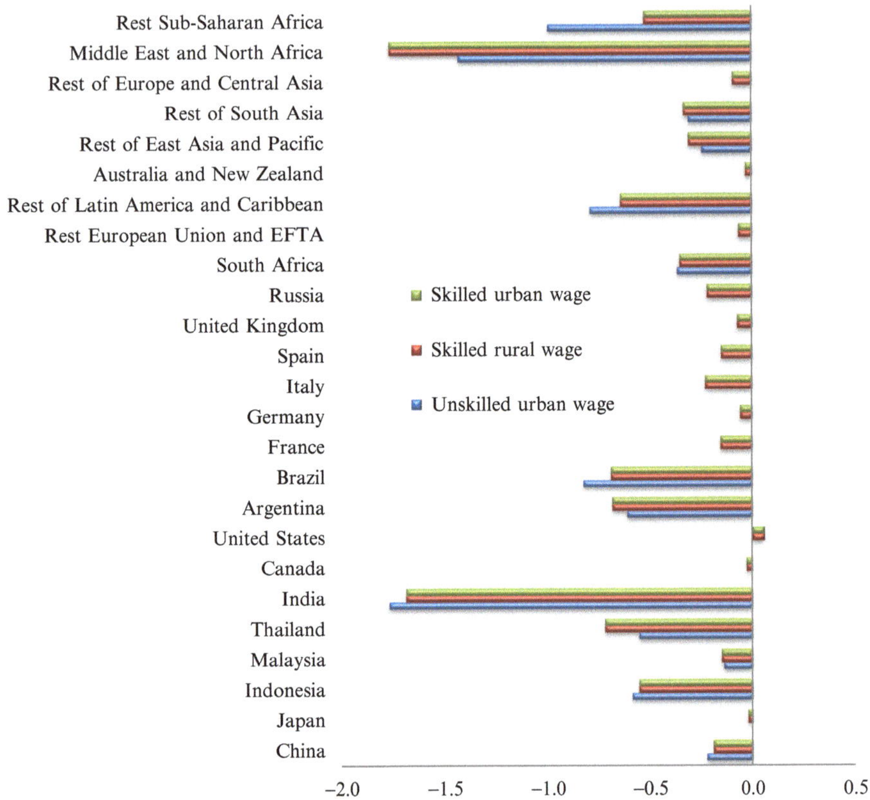

Fig. 6.4 Changes in the ratio of skilled wages to unskilled wages due to biofuel targets and mandates in 2020 (% change from BAU)

movement towards the rural areas. The movement is highest in India and Middle East and North Africa. There are also noticeable similar labor movements in Brazil, Argentina, Thailand, and Indonesia. There is no such movement of unskilled labor in developed countries.

6.6 Effects on Poverty and Income Distribution

The income and price effects from CGE were utilized in the household simulation model to analyze income distribution and poverty effects of increased biofuel production. In the poverty analysis, two poverty threshold levels were applied: $1.25 per day and $2.50 per day. The results are analyzed using the change in the GINI coefficient and the poverty headcount between 2005 and 2020.

The global income redistributions result in a composite −0.1 change to the GINI coefficient. Sub-Saharan Africa adjusts by −0.2, indicating that regional incomes

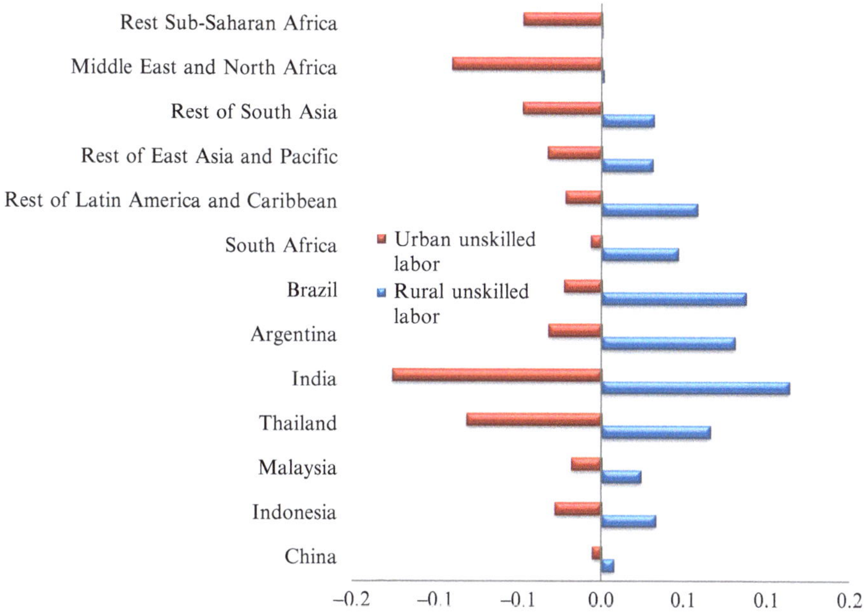

Fig. 6.5 Movement of unskilled labor due to biofuel mandates and targets in 2020 (% change from BAU)

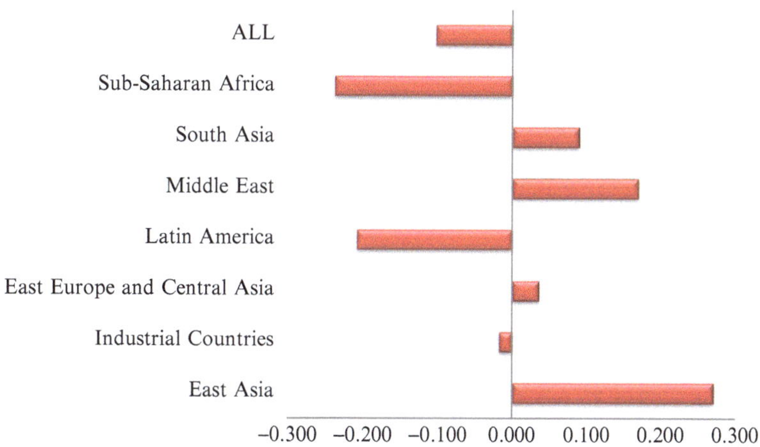

Fig. 6.6 Change in the GINI coefficient due to biofuel mandates and targets (change from BAU)

are becoming less equal (Fig. 6.6). Meanwhile, East Asia gains greater income equality by 0.275 due to the expansion of biofuels as a result of targets and mandates. Other regions that gain equality are Middle East, South Asia, and Eastern Europe and Central Asia.

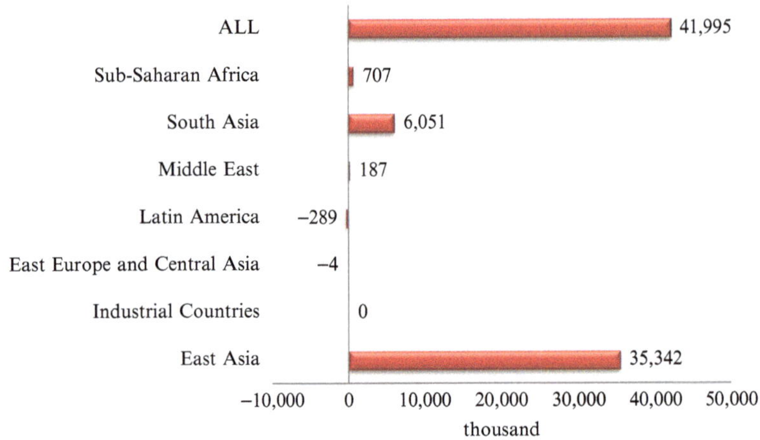

Fig. 6.7 Poverty headcount due to biofuel mandates and targets (change from BAU, US$1.25/day)

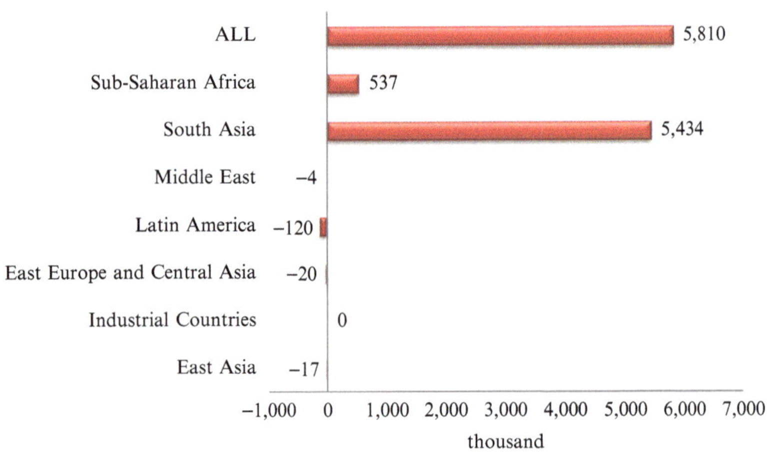

Fig. 6.8 Poverty headcount due to biofuel mandates and targets (change from BAU, US$2.50/day)

Using the US$1.25 per day threshold, the poverty headcount increases by 5.81 million around the world (Fig. 6.7). The vast majority of these new poor (5.434 million) are located in South Asia. Another 0.537 million are found in Sub-Saharan Africa. Other regions—agricultural exporters such as Latin America, Eastern Europe, and East Asia—experience reduced poverty headcount. Globally, the poverty headcount reaches 6.849 million if targeted biofuel levels are expanded at enhanced rates until 2020.

A higher poverty threshold of US$2.50/day would increase the number of poor significantly (Fig. 6.8). In East Asia, there will be 34.7 million more poor people under the TS biofuel scenario. There will be additional 6.051 million in South Asia and 0.707 million Sub-Saharan Africans under the same scenario. Table 6.2 presents a more detailed poverty effect at the country level.

Table 6.2 Impacts on poverty due to biofuel targets and mandates (change from the BaU)

Unit: Thousand	$1.25 Criteria	$2.5 Criteria		$1.25 Criteria	$2.5 Criteria
Sub-Saharan Africa	537	707	East Asia	−17	35,342
Comoros	0	0	China	0	35,684
Lesotho	0	0	Mongolia	0	0
Malawi	0	0	Malaysia	0	0
Niger	0	0	Papua New Guinea	0	0
Rwanda	0	0	Indonesia	0	−206
Sierra Leone	0	0	Cambodia	−13	−3
Zambia	0	0	Philippines	0	−27
Burundi	8	45	Thailand	−4	−23
Benin	8	33	Vietnam	0	−83
Burkina Faso	19	36			
Côte d'Ivoire	7	23	Latin America	−120	−289
Cameroon	7	38	Bolivia	−1	0
Ghana	35	25	Brazil	−34	−182
Guinea	9	13	Chile	0	−1
Kenya	0	28	Colombia	−7	−7
Madagascar	31	44	Costa Rica	0	0
Mali	14	41	Dominican Republic	0	−5
Mauritania	3	2	Ecuador	1	0
Nigeria	314	294	Guatemala	−64	0
Senegal	20	60	Guyana	0	0
Tanzania	57	55	Honduras	0	−6
Uganda	6	5	Haiti	0	0
South Africa	0	−35	Jamaica	0	−2
			Mexico	0	0
East Europe and Central	−20	−4	Nicaragua	−4	4
Bosnia and Herzegovina	0	0	Panama	−5	0
Czech Republic	0	0	Peru	−3	−11
Slovak Republic	0	0	Paraguay	−2	−2
Turkmenistan	0	0	El Salvador	−1	−76
Albania	0	0	Venezuela, Rep. Bol.	0	0
Armenia	0	0			
Azerbaijan	−1	−3	Middle East	−4	187
Bulgaria	0	0	Egypt	0	0
Estonia	0	0	Iran, I.R. of	0	0
Georgia	0	0	Tunisia	0	0
Hungary	−11	1	Jordan	0	0
Kazakhstan	0	0	Morocco	0	−20
Kyrgyz Republic	0	3	Yemen, Republic of	−4	207
Lithuania	−7	0			
Moldova	−1	0	South Asia	5,434	6,051
Macedonia, FYR	0	0	Bangladesh	−11	13
Poland	0	0	India	5,383	6,001
Romania	0	−2	Sri Lanka	−1	−8
Russia	0	0	Nepal	−8	28
Tajikistan	0	−4	Pakistan	71	17
Turkey	0	0			
Ukraine	0	0			
Uzbekistan	0	0			

Source: Cororaton and Timilsina (2012)

6.7 Closing Remarks

Expansion of the biofuel industry affects and is affected by its competing industries of fuel and food commodities. Consequently, the industry linkages and feedback effects will favor certain population over others. With more than 40 countries that have pledged biofuel targets, a large-scale expansion of biofuels would pressure food supply and food prices. Gains by biofuel industry and recession of the agricultural industry will relegate higher returns to the rural unskilled workers of the world, especially in developing nations. Even so, on a regional level, inequality and poverty abound in the already destitute, vulnerable areas. The more stricken areas tend to be regions with more ambitious goals and less biofuel facilities in place: India, Sub-Saharan Africa, Middle East and Northern Africa, Russia, and China. Conversely, countries with already advanced biofuel industry (e.g. Brazil, Argentina, Thailand, Indonesia) will experience gains in GDP per capita from 2009 to 2020, under TS and ES.

Large-scale expansion of biofuels will result in an additional 5.8 million people below the US$1.25/day poverty line by 2020, if announced biofuel targets are met. A shocking 42 million people will reach incomes below US$2.50/day. The GINI coefficient shows that East-Asian and South-Asian countries will grow poorer en masse, while Sub-Saharan Africa will experience greater distributed inequality, as well as recession.

References

Ackah, C., Bussolo, M., de Hoyos, R., & Medvedev, D. (2008). *A new dataset on global income distribution*. Washington, DC: World Bank.

Arndt, C., Benfica, R. Tarp, F. Thurlow, J., & Uaiene, R. (2008). *Biofuels, poverty and growth: A computable general equilibrium analysis of Mozambique* (IFPRI Discussion Paper 803). Washington, D.C. International Food Policy Research Institute.

Boccanfuso, D, Coulibaly, M., Timilsina, G., & Savard, L. (2013). *Macroeconomic and distributional impacts of Jatropha-based biodiesel in Mali* (World Bank Policy Research Working Paper, WPS 6500). Washington, DC: World Bank.

Cororaton, C. B., & Timilsina, G. R. (2012). *Impacts of large-scale expansion of biofuels on global poverty and income distribution* (World Bank Policy Research Working Paper, WPS 6078). *Washington*, DC: World Bank.

deHoyos, R., & Medvedev, D. (2011). Poverty effects of higher food prices: A global perspective. *Review of Development Economics, 15*(3), 387–402.

Food and Agricultural Organization (FAO). (2008). *Soaring food prices: Facts, perspective, impacts and action required. High-level conference on world food security*. Rome: Author.

Hertel, T. (2009). *Analyzing the poverty impacts of biofuel mandates*. Transition to a bioeconomy: global trade and policy issues conference, Washington DC.

Hochman, G., Rajagopal, D., Timilsina, G., & Zilberman, D. (2011). *Quantifying the causes of the global food commodity price crisis*. World Bank Policy Research Working Paper No. 5744, 7(52).

International Food Policy Research Institute (IFPRI). (2008). *Higher food prices: The what, who, and how proposed action* (IFPRI Policy Brief, IFPRI). Washington, DC: Author.

Ivanic, M., & Martin, W. (2008). *Implications for higher global food prices for poverty in low income countries*. Development Research Group, Trade Group. World Bank, Washington, DC

Keyzer, M., Merbis, M., Nubé, M., & van Wesenbeeck, C. (2008). *Food, feed, and fuel: When competition starts to bite*. SOW-VU Brief, Center for Food Studies, VU University, Amsterdam.
Runge, C., & Senauer, B. (2007). How biofuels could starve the poor. *Foreign Affairs, 86*(3), 41–53.
Timilsina, G. R., Beghin, J. C., van der Mensbrugghe, D., & Mevel, S. (2012). The impacts of biofuels targets on land-use change and food supply: A global CGE assessment. *Agricultural Economics, 43*(3), 313–330.
Mintz-Habib, N. (2010). Biofuels investment in Tanzania: Omissions in implementation. *Energy Policy, 38*(8), 3985–3997.
Mitchell, D. (2008, June). *A note on rising food prices*. Development Prospects Group, World Bank: Washington, DC.

Chapter 7
Land-Use Change and Food Supply

Jevgenijs Steinbuks and Govinda R. Timilsina

7.1 Introduction

Large-scale deployment of biofuels has a profound effect on allocation of land resources. The expansion of biofuels industry requires a greater amount of crop lands for producing biofuel feedstocks. This additional crop lands could be supplied through (1) reallocation of existing crop lands from other crops (e.g., rice, fruits and vegetables, tobacco, cotton) towards production of biofuel feedstocks (e.g., corn, sugarcane, jatropha, rapeseed), (2) conversion of forest and pasture lands to crop lands. The land-use change thus occurs directly and indirectly. For example, when forest land is converted to produce sugarcane, such conversion is termed as *direct* land-use change. When biofuels displace existing crop lands in one part of the world, and production of food crops increases in other parts of the world (e.g., by converting forest lands to crop lands), this conversion is termed as *indirect* land-use change.

Starting from pioneering works of Fargione et al. (2008) and Searchinger et al. (2008), a large number of studies have examined the impact of expansion of biofuels on land-use change at national, regional, and global levels (Al-Riffai et al. 2010; Banse et al. 2008; Dicks et al. 2009; Fabiosa et al. 2010; Gurgel et al. 2007; Hertel et al. 2010, 2013; Lotze-Campen et al. 2010; Melillo et al. 2009). These studies use increasingly sophisticated partial and general equilibrium modeling tools to examine the impacts of biofuels on land-use change. Khanna and Crago (2012) presents an exhaustive review of literature assessing land-use change impacts of biofuels. This chapter will therefore not attempt to summarize all of the forgoing work on

J. Steinbuks • G.R. Timilsina (✉)
The World Bank, 1818 H Street, N.W., Washington, DC 20433, USA
e-mail: jsteinbuks@worldbank.org; gtimilsina@worldbank.org

G.R. Timilsina and D. Zilberman (eds.), *The Impacts of Biofuels on the Economy, Environment, and Poverty*, Natural Resource Management and Policy 41,
DOI 10.1007/978-1-4939-0518-8_7,

biofuels and land-use change. Instead, it employs two novel and distinct frameworks to discuss some short- and long-term implications on land use of meeting biofuel mandates and targets announced by 40 plus countries around the world.

We start with the discussion of short-term land-use impacts based on Timilsina et al. (2012), which analyzes land-use impacts of meeting biofuel blending mandates and targets in the near decades using a global, multi-sector, multi-region computable general equilibrium (CGE) model. The study focuses on first generation biofuel technology given the dim perspectives of introducing second generation biofuels in the near decades (NRC 2011). We then proceed with long-run economic assessment, extending the work of Hertel et al. (2013) using FABLE, a dynamic optimization partial equilibrium model for the world's land resources over the next century (Steinbuks and Hertel 2012). The model solves for the intertemporal paths of alternative land uses which together maximize global economic welfare. Alternative land uses incorporated into the model include: food crops, livestock feed, pasture lands, protected natural lands, managed (commercially exploited) forests, unmanaged forests, and first and second generation liquid biofuels.

7.2 Land-Use Change due to Biofuels in the Short Run

This section discusses near-term (by 2020) impacts of biofuels on land-use change based on Timilsina et al. (2012). The baseline and the scenarios simulated in Timilsina et al. (2012) are the same as presented in Chap. 5 of this book, where economic impacts of those scenarios were discussed. The model is a global, multi-country, multi-sector CGE model. The detailed description of the model is available in Timilsina et al. (2012) or Timilsina et al. (2011). In the model, the baseline projects economic development, population growth, and biofuel production at a "business as usual" rate since 2004. Key exogenous variables in this and most other dynamic CGE models include:

- Labor supply is determined through exogenous population growth and the fixed ratio between working age population and total population.
- Productivity growth (total factor productivity) is exogenous.
- Exogenous energy price growth.[1]

The *biofuels mandate and target* scenario considers the implementation of biofuel targets by acceding nations. The CGE model computes which and where biofuel feedstocks should grow most efficiently so as to reallocate land for optimal returns. Different countries have set different years for meeting their targets by 2020.

[1] A module that can represent both conventional and unconventional oil and gas reserves and production would be ideal; however, the model used here does not have that capacity. Hence, we used energy price forecasts from other sources instead of generating them endogenously in the baseline.

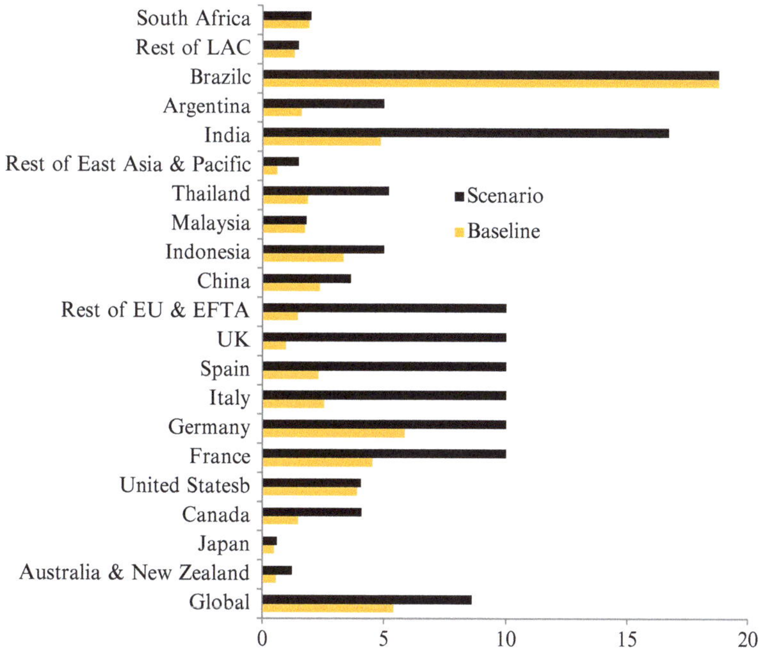

Fig. 7.1 Biofuel penetrations in the baseline and the target scenario. *Source*: Timilsina et al. (2012)

Countries that have target dates prior to 2020 are assumed to maintain the targets once they meet. In other words, once percentage targets are reached, the shares remain constant but the physical volumes change as the total transportation energy consumption increases over time. Figure 7.1 presents projected penetration of biofuels, defined as the share of biofuels in the total liquid fuel for road transportation on energy equivalent basis, for various countries and regions. In the baseline, the global penetration of biofuels, which is roughly 3 % at present, is expected to reach 5.4 % in 2020 due to policies already in place before 2009 and due to increasing oil prices. If the targets announced by all countries are realized by 2020, the global penetration of biofuels is expected to reach about 9 % by 2020.

Figure 7.2 summarizes aggregate land allocation under the biofuels mandate and target scenario for 2020 as compared to that happened otherwise in the baseline. As the figure demonstrates, land use shifts away from pasture, forestry, and non-feedstock crops towards crops that serve as feedstock for biofuels. Most of the conversion to crop land comes at the expense of forests and pastures. In aggregate, global forests recede 0.2 % and pastures decline by 0.2 % under the biofuel expansion scenario. The agricultural boom and deforestation effects are more pronounced in EU countries, Thailand, South Africa, India, and Brazil. The highest rate of deforestation due to land conversion is found in France. France also substitutes rice cultivated lands in favor of sugar and oilseed production. The high income nations make more rapid conversion of lands than do middle and lower income nations.

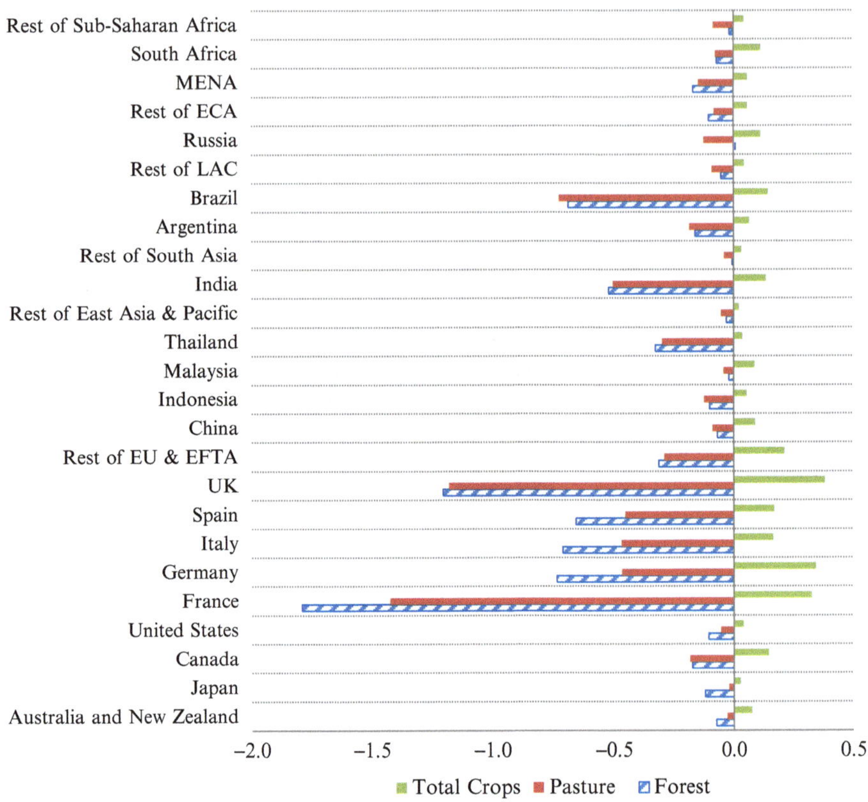

Fig. 7.2 Change in land Supply due to biofuel targets relative to the baseline in 2020 (%)

Due to market price signals, sugar crops expand the most in production around the world, while rice tends to contract in production. For more details on deforestation and land conversion, please refer to Fig. A.1 and Table A.1 in the Appendix to this chapter.

The model predicts additional diversion of lands within the crop category, especially from rice and fruits and vegetables and other non-biofuel feed stocks to sugar crops and other biofuel feed stocks (see Fig. 7.3). Although the price signals favor the production of sugar and coarse grains, other sectors lose in demand for agricultural goods. The changes in world food supply consequent to the reallocation of land uses are illustrated in Fig. 7.4. Globally, more than US$6 billion worth of food supply would decrease as compared to baseline in 2020. However, in percentage terms the global loss is relatively small. The impacts on food supply are significant in regions like India and Sub-Saharan Africa where food deficit is a persistent problem. Growing crop prices and reductions in food supply render greater vulnerability to nutritional needs in several countries, especially among their indigent populations.

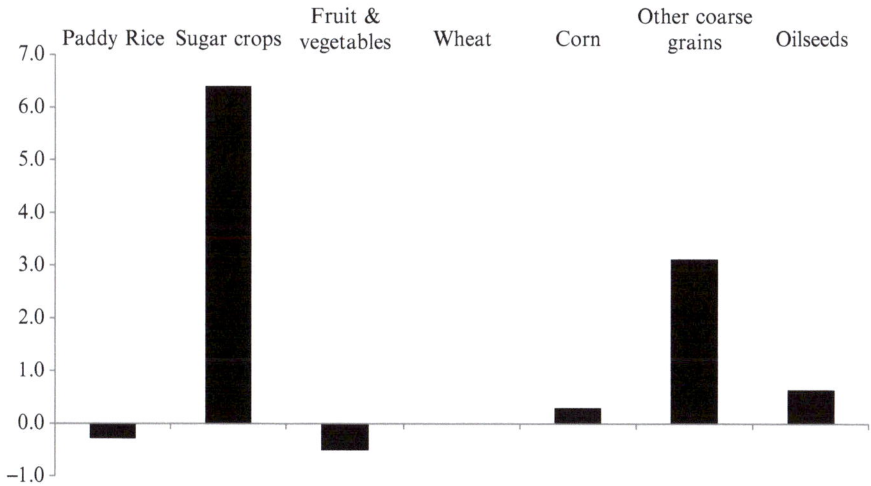

Fig. 7.3 Change in crop land supply due to biofuel targets relative to the baseline in 2020 (%). *Source*: Timilsina et al. (2012)

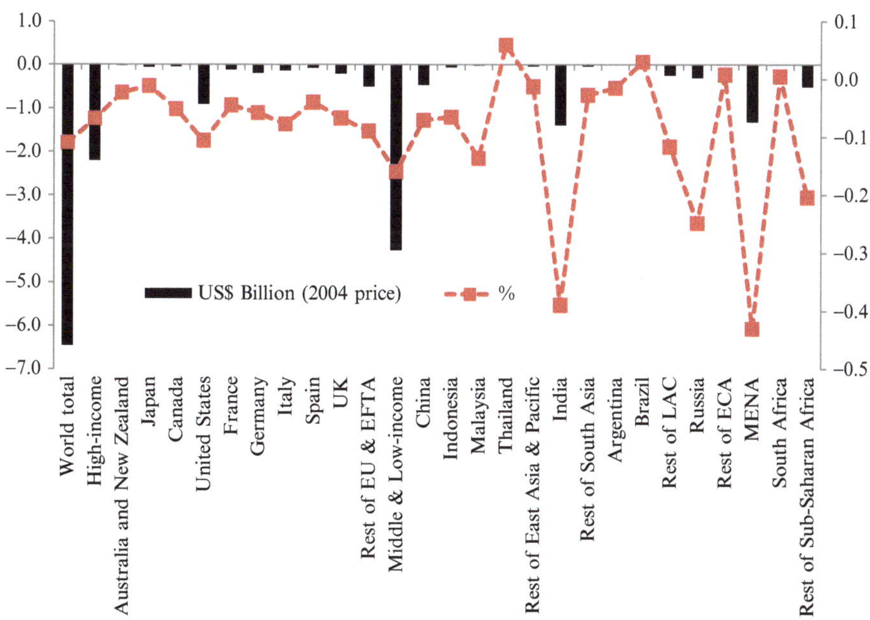

Fig. 7.4 Change in food supply due to biofuel targets relative to the baseline in 2020

7.3 Land-Use Change due to Biofuels in the Long Run

In the preceding CGE analysis of biofuels, the biofuels' growth in near decades was mainly driven by policies, such as subsidies and government mandates. However, in an environment of constrained budgets and slower economic growth, the long-run prospects for biofuels are likely to hinge on their economic and environmental contributions to global well-being. Biofuels could be attractive in the environment of high energy prices, and advances in both agricultural yields and cellulosic conversion technology for producing drop-in biofuels. In such circumstances, biofuels can have a potential to displace petroleum products and to reduce the Greenhouse Gas (GHG) emissions associated with combustion of liquid fuels (NRC 2011). For these reasons, it is useful to explore the *optimal* path of global land use for biofuels over the next century, accounting for key drivers such as increasing oil prices and potential GHG emission targets, as well as potential changes in technology and evolving consumer preferences for food, fuel, and biodiversity. Such an analysis offers a valuable guide to how global land use will be impacted by biofuels in the very long run.

The results from this section are drawn from FABLE (Forest, Agriculture, and Biofuels in a Land-use model with Environmental services), a dynamic optimization partial equilibrium model for the world's land resources over the next century (Steinbuks and Hertel 2012). The model solves for the intertemporal paths of alternative land uses which together maximize global economic welfare, potentially subject to a constraint on global GHG emissions. Alternative land uses incorporated into the model include: food crops, animal feed, pasture lands, protected natural lands, managed (commercially exploited) forests, unmanaged forests, and first and second generation liquid biofuels.

Key exogenous drivers include:

- Population growth which we assume will plateau at ten million people by 2100
- Global per capita income which rises at a rate of 2.25 %/year
- Oil prices which are assumed to rise at about 0.9 %/year over the twenty-first century
- Technological progress in the agriculture, forestry, energy, and recreation sectors
- Yields in agriculture grow linearly over most of the century, flattening when getting closer to their potential (Cassman et al. 2010)
- Energy efficiency, which grows at a rate of 1.6 %/year

Complete documentation of the model's structure, including equations, variables, and parameters is offered in technical documentation (Steinbuks and Hertel 2012).

Figure 7.5 shows the optimal allocation of lands between the alternative uses over this century *in the absence of binding biofuels mandates*. Protected forests expand in response to growing consumer demand for ecosystem services as households become wealthier. Cropland for food expands until 2035 due to increasing population and evolving consumption patterns, but declines thereafter as population and per capita demand growth would be slow and will be overtaken by technological progress in agriculture. Improvements in crop technology and agricultural yields

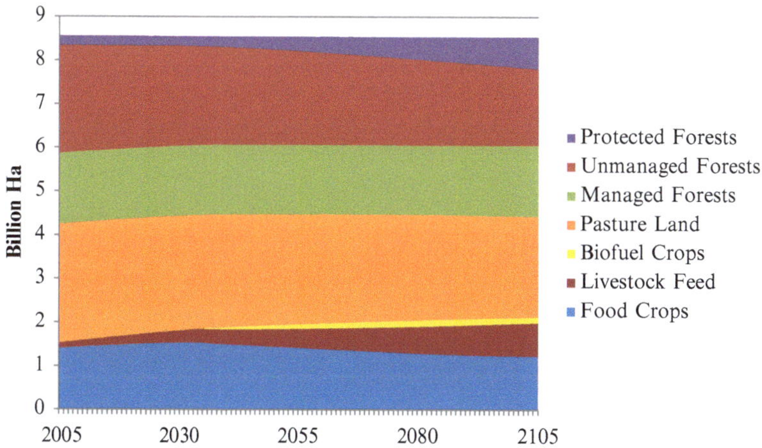

Fig. 7.5 Optimal allocation of global land resources, 2005–2105. *Source*: Steinbuks and Hertel (2012)

result in greater intensification of livestock production. As a result the area dedicated to animal feed expands considerably, whereas the pasture land declines over the course of next century. Managed forest area would change a little. Land devoted to biofuels expands steadily—particularly after second generation biofuels become commercially competitive in 2035.

Thus, even without subsidies, GHG targets, or biofuels mandates, our baseline does suggest that, if oil prices continue to grow (0.9 %/year) throughout the century, the globally optimal land area devoted to biofuel feed stocks would amount to about 150 Mha by the end of the century and biofuels would account for about 30 % of global liquid fuel consumption—mostly from second generation, drop-in biofuels at the end of the century. Of course this result is quite sensitive to the path of oil prices (Steinbuks and Hertel 2013).

As seen above, in the context of the dynamic-recursive CGE analysis, policies aimed at boosting deployment of biofuels can have a significant impact on biofuel production and global land use in the near decades. Accordingly, we wish to explore, within the context of this forward-looking model, the comparative dynamic impacts of a global biofuels mandate on global land use. We target an 8 % share of first- and second-generation biofuels in total liquid fuel consumption, which corresponds to predicted result from the CGE model if all current biofuel mandates and targets are implemented. This fully binding mandate is announced in advance, and is introduced in 2020. The consequences for land use change are shown in Fig. 7.6.

As expected, implementation of this biofuels mandate leads to increased supply of biofuels crops, and decline in land areas dedicated to food crops and pasture land. The optimal path of forest lands is largely unaffected by the mandate. With the binding biofuels mandate, the second generation biofuels will enter the market as early as 2020, and require additional 17 million hectares of land. Overall, areas dedicated to non-biofuels crops decline by about 1 % in 2020, as compared to baseline scenario. However, the impact of the biofuels mandate is relatively short lived.

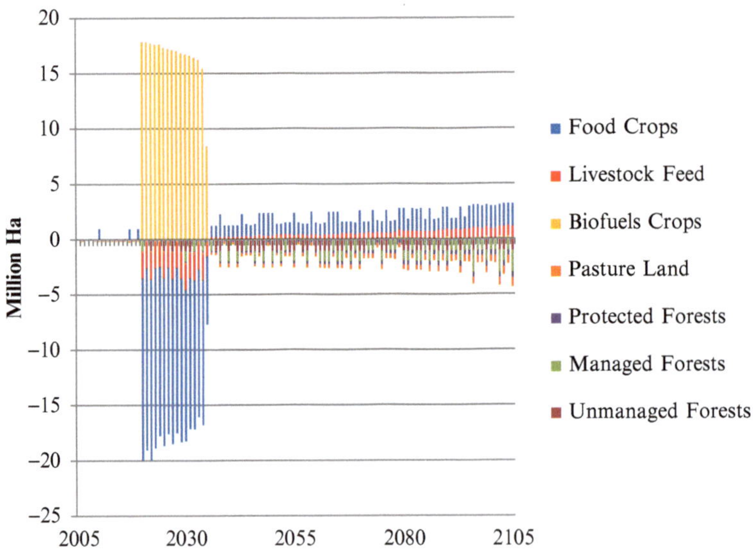

Fig. 7.6 Change in land use relative to the baseline in 2005–2105

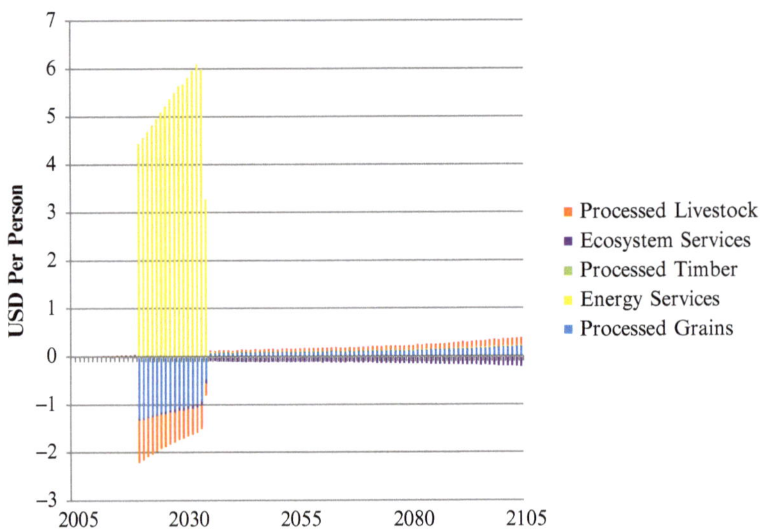

Fig. 7.7 Change in per capita land-based goods and services relative to the baseline in 2005–2105

As energy prices continue to increase over the course of this century, the biofuels mandate becomes slack in 2035. Introduction of biofuels mandate thus has very small effect on the optimal path of global land use in the long term.

Figure 7.7 shows implications of global biofuels mandate on the consumption side. Increased competition for land resources translates into reduced consumption of services from processed grains and livestock, which cumulatively fall by about $2 per person (about $13 billion at 2004 population level) in 2020. As explained above, this reduction in food supply is caused by diversion of land resources from food crops, animal feed, and pastures to biofuels feed stocks. The global consumption of land-based energy services increases over the period 2020–2035 compared to baseline scenario, reaching its maximum of $6 per person (about $38 billion at 2004 population level). When the biofuel mandate becomes slack in 2035, the consumption of land-based goods and services is little changed.

7.4 Conclusions

This chapter employs two different modeling approaches to demonstrate complex interactions between forest, pasture, and crops that affect allocation of global land use in the context of large-scale deployment of biofuels. We first show the results from the recursive-dynamic CGE model, aimed at investigating land-use implications of biofuels deployment in the near decades. Under the scenario of meeting biofuel targets and mandates announced by 40 plus countries around the world, rapid expansion of biofuels leads to increased deforestation and conversion of pasture lands in many countries. This expansion also causes diversion of lands from other food crops (e.g., rice, fruits, and vegetables) to those used for biofuels (sugar crops, corn). While planned biofuel targets are not expected to significantly affect global aggregate food supply, national food supplies would suffer in the near decades, especially for developing and poverty-stricken countries and regions, such as Sub-Saharan Africa and India.

We then proceed with the analysis of biofuels deployment in the long run, using a dynamic, forward-looking partial equilibrium model and found that even without subsidies, aggressive climate policies, or biofuels mandates, there will be a significant expansion in the globally optimal land area devoted to biofuel feed stocks. Our baseline does suggest that, if oil prices continue to grow throughout the century, land areas dedicated to biofuels feedstock would amount to about 150 Mha by the end of the century. And biofuels would account for about 30 % of global liquid fuel consumption, at the end of the century, when second generation, drop-in biofuels become competitive. Along this optimal path of global land use, biofuels mandate have a very small effect on global land use and consumption of land-based goods and services in the long term. Of course this result is quite sensitive to the path of oil prices.

Appendix

Table A.1 Deforestation due to global expansion of biofuels—% change from the baseline in 2020

Country/Regions	Forest	Pasture	Rice	Sugar crops	Fruits and vegetables	Wheat	Corn	Other grains	Oilseeds
World total	−0.2	−0.2	−0.3	6.4	−0.5	0	0.3	3.1	0.6
High-income	−0.4	−0.3	−0.2	6.3	−0.7	0.4	−0.2	7.8	3
Australia and New Zealand	−0.1	0	−0.1	0	0	0.5	0.6	0.4	1.2
Japan	−0.1	0	−0.2	0.1	0	0.4	0	0.5	0.3
Canada	−0.2	−0.2	0	−0.2	−0.3	0	2.7	0.2	0.4
United States	−0.1	−0.1	−0.1	−0.1	−0.1	0.5	0.3	0.3	0.5
France	−1.8	−1.4	−3.2	37	−2.8	0.8	−2.8	3.1	17.3
Germany	−0.7	−0.5	0	−0.4	−0.9	3.3	−0.8	3.9	11
Italy	−0.7	−0.5	−0.7	−0.8	−0.8	−0.4	−1.4	−0.4	8.6
Spain	−0.7	−0.5	−1	−0.6	−0.8	−0.1	−1.1	25.4	0.5
UK	−1.2	−1.2	0	10.8	−2.1	−4.3	0	40	4.9
Rest of EU and EFTA	−0.3	−0.3	−0.3	1.4	−0.5	0.7	−1.4	6.2	5
Middle and low-income	−0.1	−0.2	−0.3	6.5	−0.3	−0.3	0.7	−0.1	−0.3
China	−0.1	−0.1	−0.1	0	−0.2	0.4	3.3	−0.7	−0.6
Indonesia	−0.1	−0.1	−0.1	6.3	−0.2	0	−0.1	0	0
Malaysia	0	0	0.1	0.1	0	0	0	1.1	0.2
Thailand	−0.3	−0.3	−0.7	15.2	−0.6	0	0.2	−0.3	−0.6
Rest of East Asia	0	−0.1	−0.1	2.2	0	0.2	0.7	0.3	0.2
India	−0.5	−0.5	−0.9	10	−1.1	−0.7	−0.7	−0.6	−0.8
Rest of South Asia	0	0	0	0.6	0	0.1	0.1	0.1	0.8
Argentina	−0.2	−0.2	0	−0.4	−0.2	0	0.8	−0.2	0
Brazil	−0.7	−0.7	−1.1	10.7	−1.3	−1.3	−1.5	−1.1	0.4
Rest of LAC	−0.1	−0.1	−0.1	0	0	0.2	0	−0.1	1.2
Russia	0	−0.1	−0.1	0	0.1	0	0.1	0.2	1.1
Rest of ECA	−0.1	−0.1	−0.1	0.1	0	0.2	0.2	0.3	1.1
MENA	−0.2	−0.1	−0.2	0.2	0	0.5	0.7	0.3	0.7
South Africa	−0.1	−0.1	0	0.2	0.1	1	0	0.7	0.2
Rest of Sub-Saharan	0	−0.1	0	0.1	0.1	0.7	−0.1	−0.1	0.1

Source: Timilsina et al. (2012)

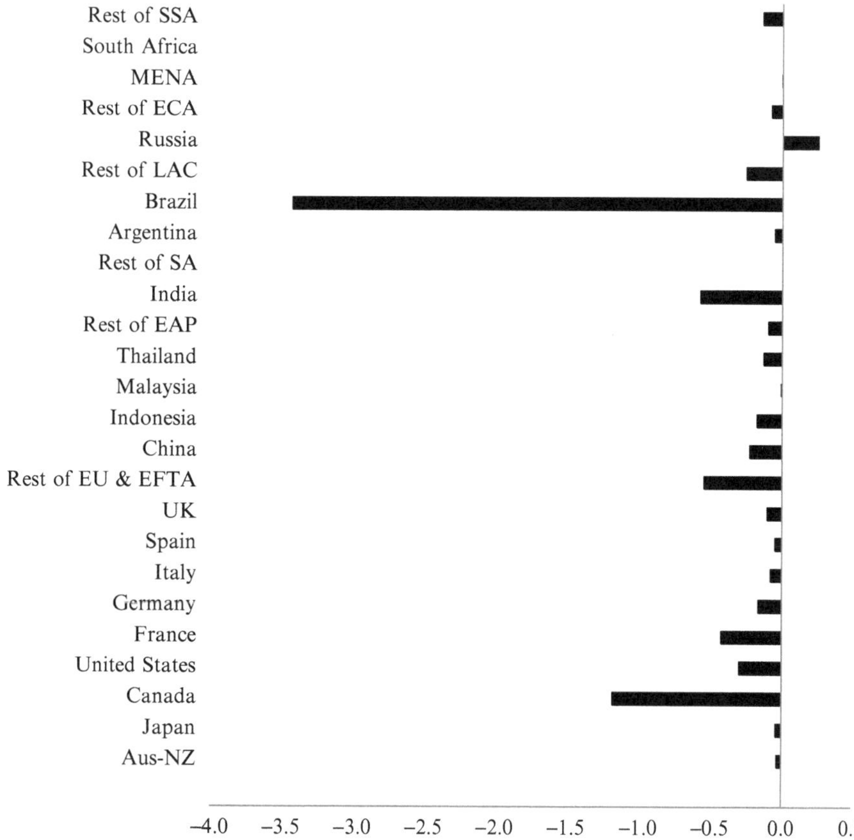

Fig. A.1 Deforestation due to global expansion of biofuels—change from the baseline in 2020 (million ha.). *Source*: Timilsina et al. (2012)

References

Al-Riffai, P., Dimaranan, B., & Laborde, D. (2010). *Global trade and environmental impact study of the EU biofuels mandate*. Washington, DC: International Food Policy Research Institute. Retrieved from http://environmentportal.in/files/biofuelsreportec.pdf

Banse, M., Meijl, H. van, Tabeau, A., & Woltjer, G. (2008). *Impact of EU biofuel policies on world agricultural and food markets*. Presented at the 107th Seminar of the European Association of Agricultural Economists, Sevilla, Spain.

Cassman, K. G., Grassini, P., & Wart, J. V. (2010). Crop yield potential, yield trends, and global food security in a changing climate. In D. Hillel & C. Rosenzweig (Eds.), *Handbook of climate change and agroecosystems* (pp. 37–51). London, England: Imperial College Press.

Dicks, M. R., Campiche, J., Ugarte, D. D. L. T., Hellwinckel, C., Bryant, H. L., & Richardson, J. W. (2009). Land use implications of expanding biofuel demand. *Journal of Agricultural and Applied Economics, 41*(02). Retrieved from http://ideas.repec.org/a/ags/joaaec/53091.html

Fabiosa, J. F., Beghin, J. C., Dong, F., Elobeid, J., Tokgoz, S., & Yu, T.-H. (2010). Land allocation effects of the global ethanol surge: Predictions from the international FAPRI model. *Land Economics, 86*(4), 687–706.

Fargione, J., Hill, J., Tilman, D., Polasky, S., & Hawthorne, P. (2008). Land clearing and the biofuel carbon debt. *Science, 319*(5867), 1235–1238.

Gurgel, A., Reilly, J. M., & Paltsev, S. (2007). Potential land use implications of a global biofuels industry. *Journal of Agricultural & Food Industrial Organization, 5*(2). Retrieved from http://ideas.repec.org/a/bpj/bjafio/v5y2007i2n9.html

Hertel, T. W., Golub, A. G., Jones, A., O'Hare, M., Plevin, R., & Kammen, D. (2010). Global land use and greenhouse gas emissions impacts of US maize ethanol: Estimating market-mediated responses. *BioScience, 3*, 223–31.

Hertel, T. W., Steinbuks, J., & Baldos, U. L. (2013). Competition for land in the global bioeconomy. *Agricultural Economics, 44*(s1), 129–138. doi:10.1111/agec.12057.

Khanna, M., & Crago, C. L. (2012). Measuring indirect land use change with biofuels: Implications for policy. *Annual Review Resource Economics, 4*(1), 161–184.

Lotze-Campen, H., Popp, A., Beringer, T., Müller, C., Bondeau, A., Rost, S., et al. (2010). Scenarios of global bioenergy production: The trade-offs between agricultural expansion, intensification and trade. *Ecological Modelling, 221*(18), 2188–2196. doi:10.1016/j.ecolmodel.2009.10.002.

Melillo, J. M., Gurgel, A. C., Kicklighter, D. W., Reilly, J. M., Cronin, T. W., Felzer, B. S. et al. (2009). *Unintended environmental consequences of a global biofuels program* (No. 168). MIT Joint Program on the Science and Policy of Global Change.

National Research Council (NRC). (2011). *Renewable fuel standard: Potential economic and environmental effects of U.S. biofuel policy*. The National Academies Press. Retrieved from http://www.nap.edu/catalog.php?record_id=13105

Searchinger, T., Heimlich, R., Houghton, R. A., Dong, F., Elobeid, A., Fabiosa, J. et al. (2008). Use of US croplands for biofuels increases greenhouse gases through emissions from land-use change. *Science, 319*(5867), 1238–1240.

Steinbuks, J., & Hertel, T. (2012). Forest, Agriculture and biofuels in a land use model with environmental services. Center for Global Trade Analysis Working Paper. Retrieved from https://www.gtap.agecon.purdue.edu/resources/res_display.asp?RecordID=3988

Steinbuks, J., & Hertel, T. W. (2013). Energy prices will play an important role in determining global land use in the twenty first century. *Environmental Research Letters, 8*(1), 014014. doi:10.1088/1748-9326/8/1/014014.

Timilsina, G. R., Beghin, J. C., van der Mensbrugghe, D., & Mevel, S. (2012). The impacts of biofuels targets on land-use change and food supply: A global CGE assessment. *Agricultural Economics, 43*(3), 315–332.

Timilsina, G. R., Csordas, S., & Mevel, S. (2011). When does a carbon tax on fossil fuels stimulate biofuels? *Ecological Economics, 70*, 2400–2415.

Chapter 8
Oil Price and Biofuels

Govinda R. Timilsina

8.1 Introduction

Does the increasing oil price have an impact on demand for biofuels? Does an expansion of biofuels put pressure on oil price to fall down? These are the questions frequently asked by policy makers and other stakeholders. However, the answer is ambiguous. Serra and Zilberman (2013) surveyed 45 studies published during 2007–2012 period, which mostly used time series data, to analyze biofuel related price transmission. Only two out of the 45 studies found an impact of biofuels on fossil fuel prices in the long-run. Twenty of those studies found that energy prices influence agricultural commodities' prices. One can argue that since feedstock costs represent more than half of the total production costs of crop-based biofuels, energy prices would impact prices of biofuels. However, existing literature have mixed findings on the relationship between oil price rise and increased prices for agriculture commodity. While some studies, such as Ciaian and Kancs (2011), Mallory et al. (2012) show the transmission of increased energy prices to agricultural commodity prices, others studies, such as Cha and Bae (2011), Hassouneh et al. (2012) find no such relationship in the long-run. This implies that increased oil price would cause substitution of petroleum products with biofuels in the long-run thereby increasing their penetration in the global energy mix. Timilsina et al. (2011) find, in the United States for example, that around a 50 % rise in oil price in 2020 from 2009 level would increase biofuel use to the level that the country has targeted through its blending mandate.

The impacts of oil prices on biofuels are assessed using different approaches: (1) econometric approach and (2) macroeconomic, particularly CGE approach. Studies, such as Cha and Bae (2011), are the examples for the first approach, whereas Timilsina et al. (2011) is an example for the second approach. Despite a large

G.R. Timilsina (✉)
The World Bank, 1818 H Street, N.W., Washington, DC 20433, USA
e-mail: gtimilsina@worldbank.org

G.R. Timilsina and D. Zilberman (eds.), *The Impacts of Biofuels on the Economy, Environment, and Poverty*, Natural Resource Management and Policy 41,
DOI 10.1007/978-1-4939-0518-8_8, © Springer Science+Business Media New York 2014

number of econometric studies analyzing the relationship between the oil prices and biofuels market; no consensus on the effect of rising oil prices on biofuel markets can be made. Therefore, structural models such as CGE that captures inter-industry linkages and feedback, would be better suited to analyze impacts of oil prices on biofuel market. In this chapter, we discuss impacts of oil prices on biofuel demand based mainly on Timilsina et al. (2011). The discussions will be supplemented with other relevant literature. Besides the discussions on the impacts of oil prices on biofuels, the chapter also briefly highlights the impacts of large-scale expansion of biofuels on fossil energy demand and prices.

8.2 Impacts of Increased Oil Prices on Biofuels

In analyzing the impacts of increased oil prices on biofuels, Timilsina et al. (2011) developed three scenarios for oil price trajectories. The first scenario assumes a 25 % higher oil price as compared to that in the baseline scenario. The baseline scenario is similar to that described in Chap. 5 of this book. The second and third scenarios assume, respectively 50 % and 100 % higher oil prices as compared to that in the baseline. Should oil prices increase by 25 % from the baseline, it would be 92 % in 2015 and 106 % in 2020 from the 2009 level. Similarly, 100 % increase from the baseline in 2020 would mean 232 % higher from 2009 level. Figure 8.1 portray oil price trajectories in the baseline and the scenarios.

Figure 8.2 shows biofuel production change from the baseline under various scenarios for oil prices. By 2020, global production of biofuels would double more than 2009 production levels. Of this increase, middle and low income countries will contribute slightly more than half of the global biofuel output. Notably, biofuel

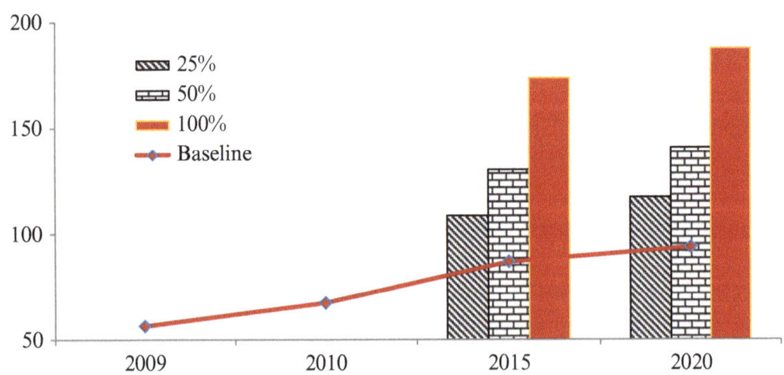

Fig. 8.1 Scenarios for oil prices (US$/barrel in 2008 price). *Source*: EIA (2009) and IEA (2009) for baseline data for period 2009–2015; for rest of the period the prices in baseline are projected by the CGE model used in Timilsina et al. (2011)

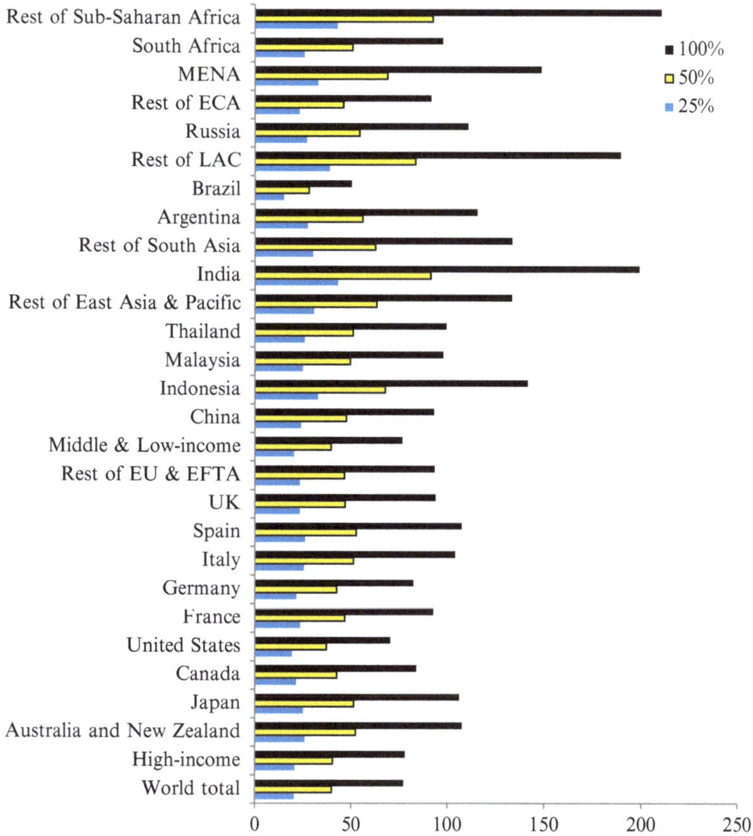

Fig. 8.2 Biofuels production in 2020 due to oil price increases. Note: EFTA stands for European Free Trade Association; LAC, ECA, and MENA refer to respectively, Latin America and Caribbean, Eastern Europe and Central Asia and Middle East and North Africa. *Source*: Timilsina et al. (2011)

production is highly sensitive to oil price changes. For example, a 25 % increase in oil price from the baseline would increase global biofuel production by 20.4 % in 2020. At each simulated oil price level, the fastest growth in national biofuel industries is found in India and Sub-Saharan Africa. Indonesia and other Latin American countries follow close behind.

Even if biofuel production grows around the world due to escalating oil prices, the penetration rate of biofuels in the fuel consumption is moderate (Fig. 8.3). Globally, the 2020 level of biofuel penetration is only 5.4 % in the baseline. A 25 % increase in oil prices from the baseline enhances the global penetration to 7 % in 2020 and a 100 % increase in oil price causes the global penetration to reach 13 %. Compared with higher income countries' penetration rates, middle and lower income nations experience more rapid biofuel penetration in response to oil price

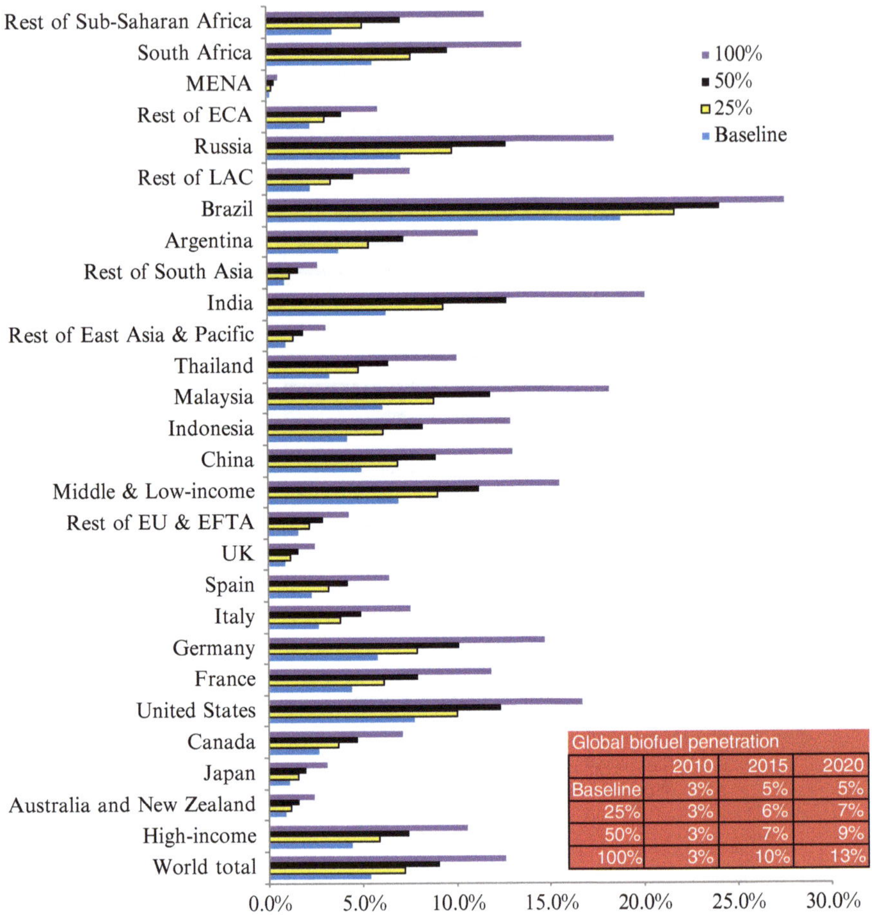

Fig. 8.3 Impacts of increased oil prices on biofuel penetration in 2020 (%). *Source*: Timilsina et al. (2011)

rise. Brazil, India, Malaysia, Russia, and the US are among the countries experiencing highest penetration rates of biofuels.

Literature employing econometric tools to investigate oil prices and biofuel markets can be divided into two groups. The first group of studies examines the relation (correlation, co-integration) between oil prices and prices of agricultural commodities. Since feedstock (i.e., agriculture commodities) is the primary raw material for biofuel production, an increased price of feedstock raises the production costs of biofuels, thereby raising their prices and reducing their demand. The second group of studies examines the relation between oil prices and biofuel markets directly (Du and McPhail 2012).

The results of econometric analysis investigating oil prices and prices of agriculture commodities vary widely across studies depending upon the technique or model used for the analysis. Studies examining long-run relationship between oil prices and prices of agriculture commodities conclude that oil prices do increase prices of agriculture commodities (see e.g., Ciaian and Kancs 2011; Nazlioglu 2011; Mallory et al. 2012). On the other hand, studies that focus on short-term relationship between oil prices and prices of agriculture commodities differ in their findings. While some studies, such as (Gilbert 2010; Esmaeili and Shokoohi 2010) find oil prices raising prices of agriculture commodities indirectly, others do not find any effect of oil price on the prices of agriculture commodities in the short-run (see e.g., Cha and Bae 2011). Existing literature do not find a direct relationship between oil prices and biofuel prices with some exceptions, such as Busse et al. 2012, which find co-integration between diesel and biodiesel and Rajcaniova and Pokrivcak (2011) that find an evidence of co-integration between prices of crude oil and ethanol.

8.3 Impacts of Biofuels on Other Energy Commodities

To date, the share of biofuels in global energy supply mix is very small. In 2010, biofuels accounted for less than 3 % of global liquid fuel consumed by road transportation (Timilsina 2013). An econometric analysis using historical data, obviously do not show an impact of biofuels on oil prices. This is also demonstrated in a number of time-series analysis on investigate relationship between prices of biofuels and oil (Du and McPhail 2012; Pokrivcak and Rajcaniova 2011). However, a study assessing impacts of large-scale expansion in future could show some impacts on fossil fuel demand and prices. For example, using a global CGE model, Timilsina et al. (2012) show that if the 40 plus countries, which have announced biofuel mandates and targets, implement policies to realize those mandates and targets, fossil demand and prices would be impacted (Fig. 8.4a, b). Demand for gasoline would drop by 1 % in 2020 from that in the baseline. Similarly, the demand for diesel would drop by 0.7 %. The global reduction in these petroleum products would cause their price to drop as well. However, the magnitude of price drop is not big because the share of biofuels in the global demand for liquid fuels in transportation would be only around 3 percentage point higher than that in the baseline.[1]

Hochman et al. (2011) also finds, using a partial equilibrium economic model, that if global demand for biofuels increases by 20 % from 2007 level, it would reduce prices of petroleum products by 2 % in oil-importing countries and 8 % in oil-exporting countries. Consequently, the global consumption of gasoline and diesel would drop by 0.4 %.

[1] The share of biofuels in 2020 was 5.4 % in the baseline, whereas it was 9 % when the announced mandates and targets were implemented. Please see Chap. 5 for more discussion.

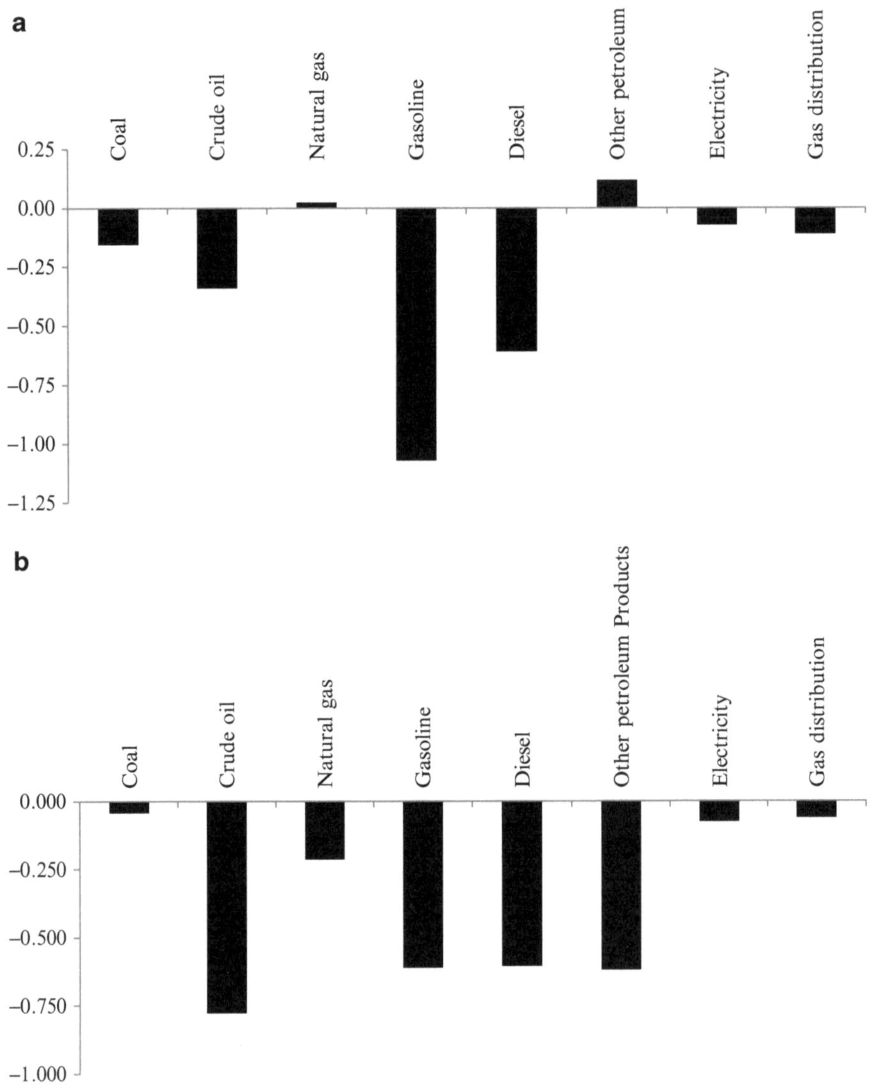

Fig. 8.4 (**a**) Impacts of biofuel expansion on the demand for other energy commodities. (**b**) Impacts of biofuel expansion on prices of other energy commodities

8.4 Conclusions

The relationship between oil price and biofuels has been investigated in the literature. However, the findings are mixed specially of econometric analysis. While most econometric literature indicate that oil prices affect biofuel markets, the magnitude of effects are not known well. Based on Timilsina et al. (2011), which quantifies, using a structural model, the magnitude of oil price's impacts on biofuels, this chapter highlights the role of increased oil prices to influence global biofuel markets in the long-run.

The global biofuel market is sensitive to prices of oil. For example, a 25 % increase in oil price from the baseline causes the global production of biofuels by 20 % in 2020. Production of biofuels in countries such as India, Sub-Saharan Africa, Indonesia, and Latin America, is relatively more sensitive compared to other countries.

The reverse relationship (i.e., the impacts of expansion of biofuels on demand for and prices of other energy commodities) is rather weak. Econometric analysis using historical data obviously demonstrate that there is no impact of change in biofuel prices on prices of other energy commodities as biofuels account for very small fraction (less than 3 %) of the global liquid fuel demand for transportation. Structural model, such as used by (Timilsina et al. 2011) confirms that the relationship would remain weak in the future as well.

References

Busse, S., Brümmer, B., & Ihle, R. (2012). Price formation in the German biodiesel supply chain: A Markov-switching vector error correction modeling approach. *Agriculture Economics, 43*, 545–560.

Cha, K. S., & Bae, J. H. (2011). Dynamic impacts of high oil prices on the bioethanol and feedstock markets. *Energy Policy, 39*, 753–760.

Ciaian, P., & Kancs, A. (2011). Interdependencies in the energy–bioenergy–food price systems: A cointegration analysis. *Resource and Energy Economics, 33*, 326–348.

Du, X., & McPhail, L. (2012). Inside the black box: The rice linkage and transmission between energy and agricultural markets. *Energy Journal, 33*, 171–194.

Energy Information Administration (EIA), (2009). *International Energy Outlook*, USDOE, Washington, DC.

Esmaeili, A., & Shokoohi, Z. (2010). Assessing the effect of oil price on world food prices: Application of principal component analysis. *Energy Policy, 39*, 1022–1025.

Gilbert, C. (2010). How to understand high food prices. *Journal of Agriculture Economics, 61*, 398–425.

Hassouneh, I., Serra, T., Goodwinand, B. K., & Gila, J. M. (2012). Non-parametric and parametric modeling of biodiesel, sunflower oil, and crude oil price relationships. *Energy Economics, 34*(5), 1507–1513.

Hochman, G., Rajagopal, D., & Zilberman, D. (2011). The effect of biofuels on the international oil market. *Applied Economic Perspectives and Policy, 33*(3), 402–427.

International Energy Agency (IEA), (2009). *World Energy Outlook*, OECD, Paris.

Mallory, M., Irwin, S. H., & Hayes, D. J. (2012). How market efficiency and the theory of storage link corn and ethanol markets. *Energy Economics, 34*, 2157–2166.

Nazlioglu, S. (2011). World oil and agricultural commodity prices: Evidence from nonlinear causality. *Energy Policy, 39*, 2935–2943.

Pokrivcak, J., & Rajcaniova, M. (2011). Crude oil price variability and its impact on ethanol prices. *Agricultural Economics, 57*, 394–403.

Rajcaniova, M., & Pokrivcak, J. (2011). The impact of biofuel policies on food prices in the European Union. *Journal of Economics, 5*, 459–471.

Serra, T., & Zilberman, D. (2013). Biofuel-related price transmission literature: A review, *Energy Economics, 37*, 141–151.
Timilsina, G.R., Beghin, J., van der Mensbrugghe, D. and Mevel, S. (2012). The impacts of biofuel targets on land-use change and food supply: A global CGE assessment, *Agriculture Economics, 43*, 313–330.
Timilsina, G. R. (2013). Biofuels in the long-run global energy supply mix for transportation. *Philosophical Transaction A* of the Royal Society, December: 372–392.
Timilsina, G. R., Mevel, S., & Shrestha, A. (2011). Oil price, biofuels and food supply. *Energy Policy, 39*(12), 8098–8105.

Chapter 9
Biofuels and Climate Change Mitigation

Govinda R. Timilsina and Simon Mevel

9.1 Introduction

Many countries present climate change mitigation benefit as one of the main rationales to defend their policies to promote biofuels. However, the role of biofuels on climate change mitigation remains ambiguous. Whether or not biofuels save greenhouse gas (GHG) emissions depends on how the savings are estimated. The GHG mitigation potentials of biofuels are normally assessed through three different approaches: project level approach, life-cycle approach, and an approach that accounts for indirect land-use change (ILUC) effect. These approaches are discussed below.

Project level approach: It assigns the GHG contents of fossil fuels replaced by biofuels as their GHG savings. It does not account for the release of GHG emissions during any activities involved in the production and delivery process, such as land cultivation and transportation of final products (e.g., ethanol, biodiesel) to blending stations. The underlying assumption here is that GHG release occurred in the supply chain of biofuel production is equal to that released in the supply chain of production of fossil fuels that is replaced with biofuels. With this assessment, any type of biofuel can save GHG emissions when it replaces fossil fuels because the former is carbon neutral. A liter of ethanol produces around 67 % of energy or mileage compared to that of gasoline, implying that ethanol could save around 67 % of GHG through gasoline replacement. Similarly, biodiesel could save around 86 % of GHG through diesel replacement.

G.R. Timilsina (✉)
The World Bank, 1818 H Street, N.W., Washington, DC 20433, USA
e-mail: gtimilsina@worldbank.org

S. Mevel
United Nations Economic Commission for Africa (UNECA), RITD,
P.O. Box 3005, Addis Ababa, Ethiopia

G.R. Timilsina and D. Zilberman (eds.), *The Impacts of Biofuels on the Economy, Environment, and Poverty*, Natural Resource Management and Policy 41,
DOI 10.1007/978-1-4939-0518-8_9, © Springer Science+Business Media New York 2014

Life-cycle assessment (*LCA*): It includes change in GHG emissions throughout the supply chains of biofuels and fossil fuels to be replaced with biofuels. For biofuels, GHG released in feedstock production (including land conversion if new land is used) and transportation, as well as GHG emissions in the refinery are accounted for. Similarly, for fossil fuels, GHG emissions in upstream petroleum activities, refining, and transportation are also included. Defining the supply-chain boundary and uncertainties on carbon coefficients at various stages of the supply chain are the key constraints to lifecycle approach.[1] GHG savings estimated through a LCA approach varies substantially across projects even if the feedstock is the same. This is because of varying assumptions on system boundaries, co-products accounting, energy sources used in the production of agricultural inputs and feedstock conversion, and the type of land (existing crop lands vs. newly converted from forest or pasture) used for feedstock production. Figure 9.1a, b illustrates the variations in GHG savings from different feedstocks.

Corn ethanol causes the lowest savings of life-cycle GHG emissions, whereas miscanthus-based second generation ethanol has the highest potential to reduce GHG emissions. Second generation or cellulosic ethanol produced from miscanthus, switchgrass, and corn stover could save more than 100 % of GHG emissions as they do not only replace gasoline but also sequester CO_2 emissions from the atmosphere. In the case of biodiesel, soybean has the highest potential for GHG reduction. As illustrated in Fig. 9.1, GHG mitigation potentials of most biofuels vary widely.

ILUC: This assessment is for capturing the ILUC effects of biofuels. The ILUC effects occur as food demand is ever increasing due to population and income growth. The increased demand for food and a new demand for agricultural commodities for biofuels would increase the overall demand for agricultural commodities. While part of this increased demand could be met through yield increase, the most of it would require a new land thereby causing deforestation and conversion of pasture lands. This implies that a biofuel program or policy in a country or region could cause land conversion not only in that region or country but in other regions of the world where production is most competitive. For example, diversion of European sugarbeet for biofuels could trigger expansion of sugarcane production in Brazil, where sugarcane production is competitive, in order to maintain the supply of sugar. The conversion of lands is associated with carbon release from soil and biomass. Some soils, such as peat land in Indonesia, are highly carbon-rich. Sometimes, the indirect carbon release might be higher than the direct release due to biofuel production.

[1] While a life-cycle approach could approximate GHG savings of a biofuel project, it does not trace GHG leakage caused beyond project boundary that normally occurs when biofuel expansion is carried out at a large-scale. For example, biofuel blending mandates in the US and Europe could increase biofuel production in Brazil and Indonesia. The production might come from feedstock grown in new lands supplied through conversion of forest or pasture lands. The conversion of lands releases GHG emissions and it is referred to as indirect land use change (ILUC) effect of biofuels.

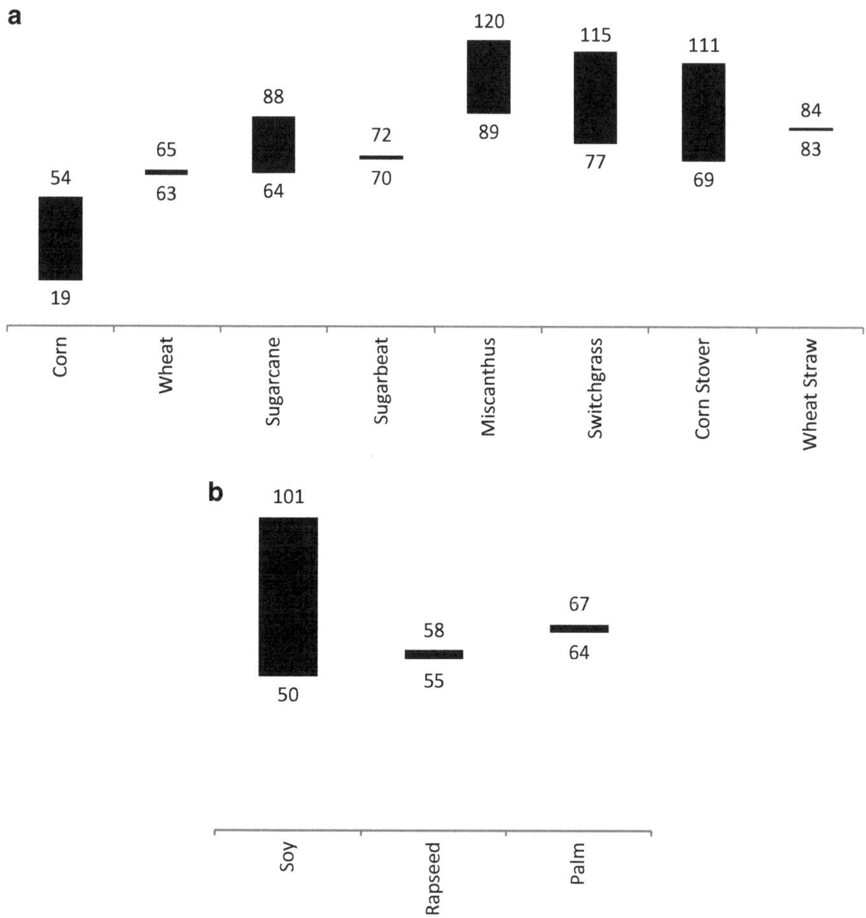

Fig. 9.1 GHG savings estimated through LCA approach. (**a**) Percentage savings of GHG emissions due to substitution of gasoline with ethanol. (**b**) Percentage savings of GHG emissions due to substitution of diesel with biodiesel. *Notes*: Calculated based on direct life-cycle emission intensities of various feedstocks compiled by Khanna and Crago (2012). Minimum value of GHG savings is calculated by subtracting maximum GHG intensity of biofuels (ethanol and biodiesel) from minimum GHG intensity of corresponding petroleum products (gasoline and diesel). Similarly, maximum value of GHG savings is calculated by subtracting minimum GHG intensity of biofuels from maximum GHG intensity of corresponding petroleum products

Two types of economic models are normally used in assessing the GHG mitigation potential of biofuels accounting for ILUCs: (1) partial equilibrium models and (2) general equilibrium models. Khanna and Crago (2012) present a good overview on the distinction of these models with examples. The first types of models are focused on agricultural sector and capture all aspects of production, consumption, and international trade of agricultural commodities and main inputs, such as fertilizers, used for production of agricultural goods and services. Since these models do

Table 9.1 Examples of studies estimating GHG savings of biofuels including ILUC effects

Study	Biofuel programs/policies	Model used	Main findings
Timilsina and Mevel (2011, 2013)	Implementation of biofuel mandates and targets announced by 40 plus countries around the world by 2020	Global CGE model	No GHG savings by completion of the program (2020); it would take 23 years after the completion of the program to realize GHG savings; if forest conversion is not allowed, GHG savings would be realized 1 year after the completion of the program
Laborde (2011)	Implementation of EU biofuel policies by 2020	MIRAGE-Biof	The ILUC effect of EU biofuels mandate eliminate more than two-thirds of the direct emission savings estimated for 2020
Fischer et al. (2009)	Meeting global biofuel targets and mandates through first and second generation biofuels	Global CGE model	If current global biofuel targets/mandates are met it would take 30–50 years to offset the GHG emissions caused by ILUC effects
Dumortier et al. (2011)	Production of 55 billion liters of corn ethanol in United States over the 30 years period	CARD model	No GHG savings; depending on scenarios and data assumptions in the model, it takes 74–137 years to offset GHG emissions caused by ethanol directly and indirectly

not necessarily include every sector and agents of the economy, they are referred to partial equilibrium models. The FAPRI-CARD model developed by the Food and Agricultural Policy Research Institute (FAPRI) and the IMPACT model developed by the International Food Policy Research Institute (IFPRI) are two good examples of partial equilibrium models used for ILUC impacts of biofuels. The second types of model are referred to as general equilibrium models. These models represent the linkage between production sectors, between production sectors and other economic agents–such as households and governments–and fully capture international trade of all goods and services. These are the most common models used for assessing ILUC impacts of biofuels. Examples of these models include Timilsina et al. (2010); Al-Riffai et al. (2010); Hertel et al. (2010); Fischer et al. (2009).

Although a large number of studies have been carried out to estimate ILUC effects, there is no consensus on any estimate. This is because of high uncertainties involved in the estimations. Different studies use different models to estimate ILUC effects; these models vary on two fronts. First, the database and underlying assumptions are different; secondly key parameters such as projection of yield, treatment co-products, assumptions about the types of land use change, and the methods for estimating GHG emissions are different.

An assessment of ILUC for a single biofuel project in isolation may not be relevant as ILUC effect of a project would be too small and too cumbersome to trace. However, it is important to measure ILUC effects of biofuel programs or policies. Normally, global macroeconomic models with explicit representation of bilateral trades have been used to assess ILUC effects of biofuels. Some examples of studies investigating climate change mitigation effects of biofuels are presented in Table 9.1.

This chapter seeks to discuss the role of biofuels in global climate change mitigation based on Timilsina and Mevel (2011, 2013). We present a comparison of greenhouse gas emissions under the baseline and the scenario of full realization of biofuel mandates and targets announced by 40 plus countries around the world by year 2020. While the baseline assumes continuation of biofuel policies already implemented before 2009, the scenario considers the biofuel mandates and targets which have been already announced but yet to be implemented. The mandates and targets will be implemented by 2020 following the schedules specified in their announcement.[2] The baseline and scenario presented here is the same as presented in Chapter V with one distinction that both baseline and scenario are projected up to 2040 to capture the carbon payback period.

9.2 Methodology to Calculate GHG Emissions

The CGE model used by Timilsina and Mevel (2011, 2013) captures GHG emissions under the baseline and biofuel expansion scenarios through the following activities:

- *Consumption of fossil fuels*: multiplying the volume of a fossil fuel consumed by a production sector and an economic agent (e.g., households, governments) by emission coefficients or carbon content of that fuel. The national emissions from fossil fuels are the sum of emissions across the fuels and across production sectors as well as economic agents. Emission coefficients are based on 2009 CO_2 emissions data from the International Energy Agency (IEA). Reduced emissions due to expansion of biofuels are calculated by subtracting CO_2 inventory from fossil fuel consumption under the scenario from that under the baseline.
- *Land use change*: GHG release due to land-use change in a given year is calculated in four steps. First, the change in carbon stock on biomass due to expansion of biofuels is calculated by subtracting carbon stock on biomass under the scenario from that under the baseline. Second, the stock is then converted to annual flow (or annual GHG release) by subtracting previous years' GHG stock change from that of the current year. Third, the annual carbon flow from change in biomass stock is multiplied by oxidization rates and carbon to carbon di-oxide ratio (3.44) to get annual CO_2 change due to land-use. Fourth, annual CO_2 release from soil carbon is added to annual CO_2 release from biomass to get total CO_2 release from land-use change. The method follows the guidelines developed by the Intergovernmental Panel on Climate Change (IPCC 2006). Relying on the IPCC Tier I approach, calculations need account for three types of biomass: (1) above ground, (2) below ground, and (3) soil. Emission stocks are documented

[2] The authors also analyze a scenario where the mandates and target are doubled to further stimulate biofuel penetration in the global energy supply mix. For results of that scenario, interested readers may refer to Timilsina and Mevel (2011) or Timilsina and Mevel (2013).

for each agro-ecological zone (AEZ).[3] Soil carbon is a flow variable accounting for emissions over the past 20 years. When a land conversion occurs, GHG release is assumed to continue over the next 20 years.

9.3 The Impacts of Biofuel Expansion on GHG Emissions

9.3.1 Impacts on Annual Emissions

As explained in the preceding section, total GHG emissions are a composite of fossil fuel emissions and carbon release from land-use change. Compared to the baseline scenario, the global GHG emissions from fossil fuel decreases over the years as the replacement of fossil fuels with biofuels continue to increase due to the mandates and targets (Fig. 9.2). The global GHG emissions due to land-use change (i.e., deforestation and cultivation of pasture lands) would decrease over years. This is because, once land conversion occurs, the same land is utilized to produce biofuel feedstock again and again. The further we go, the less new lands we need as long as biofuel mandate remains the same. Although decreasing over time, the global emissions due to land use change would be much higher than the baseline level in earlier period. The reduction in emissions from the baseline through fossil fuel replacement is not sufficient to offset the increased emissions from the baseline due to land use change, Thus, net emissions would be higher than that in the baseline. But, by

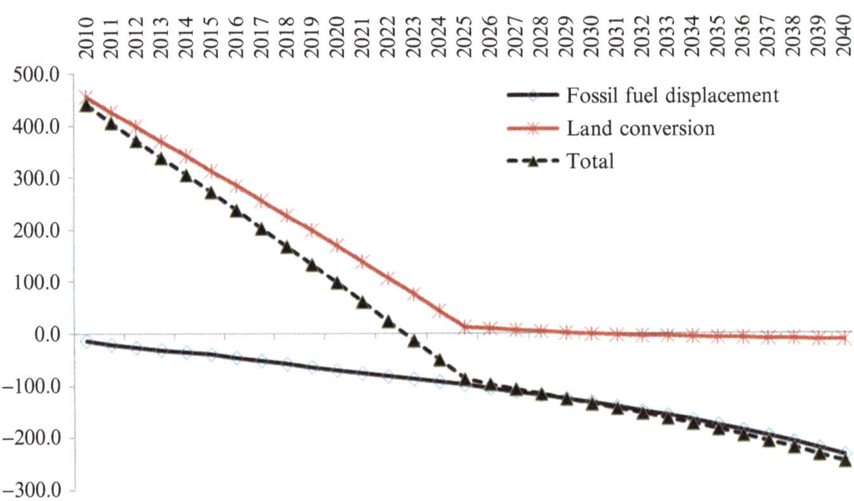

Fig. 9.2 Change in global annual CO_2 emissions from the baseline due to expansion of biofuels (million tons of CO_2)

[3] Please refer to Timilsina and Mevel (2011, 2013) for emission coefficients for various AEZs.

year 2023, the emission reduction due to fossil fuel displacement would be higher than that emission release due to land conversion thereby causing net reduction of GHG emissions. Starting 2032, there would be no more release of GHG emissions from land-use change as well, instead there would be reconversion of crop lands to pasture and forest lands thereby causing net sequestration of GHG emissions. This is because biofuel mandates and targets beyond 2020 are assumed to be kept at 2020 level. Thus, the analysis demonstrates that expansion of biofuels causes increase in global GHG emissions in the short-run but reduces GHG emissions in the long-run as long as biofuel mandates remain at the same level.

9.3.2 *Impacts on Cumulative Emissions*

Most GHG emissions from changes in land usage will incur at the time of land conversion with exception of emissions from soil carbon and harvested wood products. As GHG savings accumulate over time to offset the biomass release from land, it might be more useful to represent GHG emissions in cumulative terms, rather than in annual terms. Figure 9.3 shows the effects of meeting biofuel mandates and targets: biofuel penetration is weighed against concurrent GHG emission effects and CO_2 debt.

As illustrated in Fig. 9.3, up to 2020, the rate of biofuel penetration accelerates, causing emissions to continue rising (at a diminishing rate) and carbon debt to fall (at a diminishing rate). The global penetration of biofuels reaches 9.6 % by 2040 under the scenario to implement biofuel mandates and targets by 2020 and no incremental mandates and targets implemented thereafter. The relationship between biofuel expansion, emissions, and carbon debt all reach stabilization after 2020, as the biofuel promotional policies are held constant thereafter. The carbon debt graph shows how many years it would take to "pay off" land conversion due to biofuel promotion. It would take more than 23 years (i.e., 2043) after completion of implementing biofuel mandates (i.e., year 2020) to realize GHG savings from fossil fuels to compensate for GHG released through land conversion.

The bottom panel of Fig. 9.3 presents an interesting insight which is often ignored in the existing literature (e.g., Dumortier at al. 2011, Searchinger et al. 2008; Fargione et al. 2008). This ignorance might have resulted in heavy inflation of carbon payback periods in those studies. The insight is as follows. If an analysis is carried out for a particular year, for example, conversion of peat land in Indonesia to produce biofuels, most of the emissions release in the year of land conversion as all biomass is burned down during that year. The amount of GHG emissions would be very high. If carbon payback period is calculated by dividing this amount of emission by the amount saved through the replacement of fossil fuels, the resulted number would be very high. For example, in 2010 (the first year in Fig. 9.3), the carbon payback period is around 50 years. However, the same land is used to produce biofuel feedstocks again and again. Biofuels produced from this new feedstock also replaces fossil fuels. As we go further and further, more and more fossil fuels are replaced thereby decreasing the carbon payback period overtime.

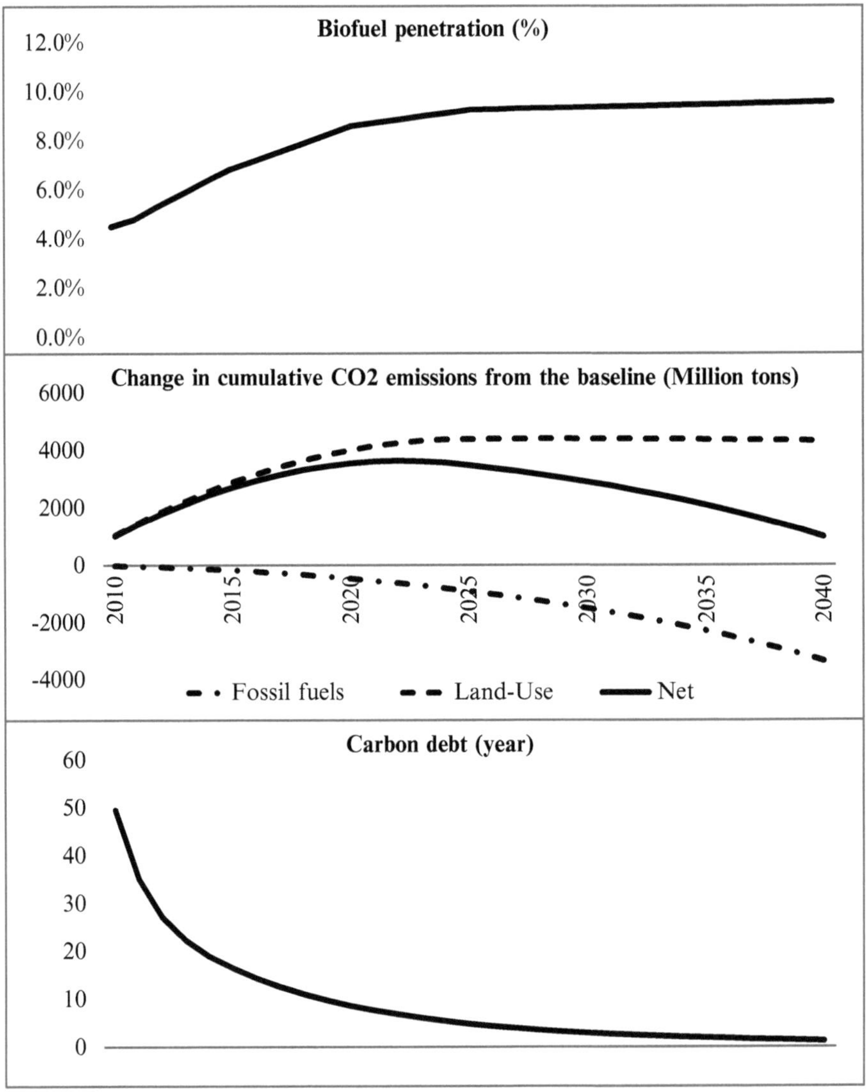

Fig. 9.3 Biofuel penetration, GHG emissions, and carbon payback period. *Source*: Timilsina and Mevel (2011, 2013)

9.4 Securing Climate Change Mitigation from Biofuels

One of the key challenges to biofuel expansion is how to limit the carbon debt that biofuel production causes. If biofuel mandates and targets are fully materialized by 2020, the global deforestation would reach about 5 million ha. that year. Table 9.2 details the deforestation impacts due to the implementation of the biofuels. Greater amounts of deforestation (i.e., million ha.) are identified for Brazil and Canada, whereas greater rates of deforestation are found in the UK, France, Thailand, and India.

Table 9.2 Change in Deforestation from the baseline due to expansion of biofuels

Country/Region	Deforestation		Deforested land as
	Million hectares	% Change from the baseline	Percentage of total pasture land
World total	4.8	0.1	0.2
High-income	2.7	0.2	0.4
Australia–NZ	0	0.1	0
Japan	0	0.1	5.8
Canada	1.3	0.2	6.2
United States	0.2	0.1	0.1
France	0.3	1.3	2.9
Germany	0.1	0.6	2.1
Italy	0.1	0.6	1.4
Spain	0	0.6	0.5
UK	0.1	1.2	1
Rest of EU and EFTA	0.5	0.3	1.6
Middle and low-income	2.2	0.1	0.1
China	0.2	0.1	0.1
Indonesia	0.2	0.1	6.8
Malaysia	0	0.1	6
Thailand	0.4	1.1	170.8
Rest of EAP	0.1	0	0.1
India	0.7	0.7	5.4
Rest of SA	0	0	0
Argentina	0	0.1	0
Brazil	1.5	0.3	0.9
Rest of LAC	0.1	0	0.1
Russia	−1.2	−0.1	−1.5
Rest of ECA	0.1	0.1	0
MENA	0	0.1	0
South Africa	0	0.1	0
Rest of SSA	0.1	0	0

Source: Timilsina and Mevel (2011, 2013)

Although Canada does not experience significant deforestation relative to its land endowment, the magnitude of its land conversion, expansion of its domestic biofuel sector, and instated import duties are substantive. In Thailand's case, land is converted from other crop uses—namely rice—to biofuel feedstock. In fact, Thailand and some other nations have forest preservation policies. Thus, biofuel feedstock demands would more likely require land conversion from pastures and other uses than forest.

The last column of Table 9.2 shows deforested lands in various countries/regions as percentage of their available pasture lands. This indicator shows whether or not a country/region has sufficient pasture lands to meet new land demands for biofuel expansion to meet the targets and mandates. At the global level, the land that comes from deforestation to meet the new land demand for biofuel expansion represents a small fraction (0.2 %) of the pasture land globally available. This indicates that biofuels expansion can be carried out without deforestation at the global level.

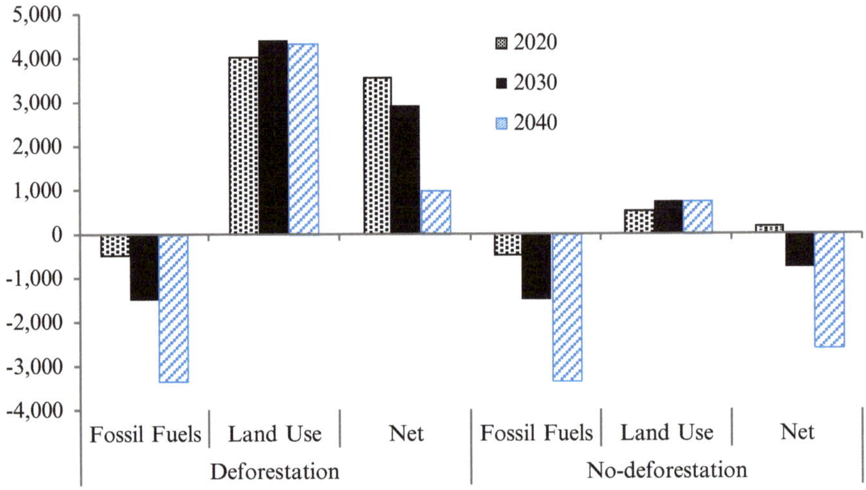

Fig. 9.4 GHG emissions under deforestation and no-deforestation cases (million tons CO_2). *Source*: Timilsina and Mevel (2011, 2013)

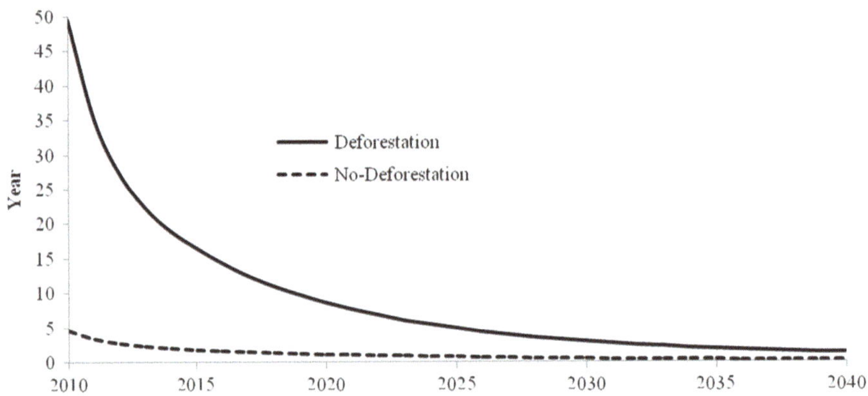

Fig. 9.5 Carbon payback periods under deforestation and no-deforestation cases. *Source*: Timilsina and Mevel (2011, 2013)

Pastures are not protected by regulation and seldom require deforestation. Thus, the carbon debt tradeoff between converting pasture land is less than for converting forests. There is some room for efficiency gains, as pastures begin to experience land-use competition and pressure. Eventually, receding pasture lands will require intensification of livestock activities and raise the cost for meats. However, there is not enough pasture lands available in some countries, particularly Thailand. In countries like Canada, Japan, Malaysia, and Indonesia, higher percentage of pasture lands are needed to avoid deforestation due to biofuel expansion.

Figures 9.4 and 9.5 extrapolate and compare GHG emissions and carbon debt under distinct sources for land conversion. Notably, if forests are converted, then deforestation causes over four billion tons of CO_2 emission in 2020. If forests are protected by regulation and pasture lands are converted to meet new land demand for biofuel expansion, net GHG release to atmosphere due to biofuels decreases by 60 folds in 2020 thereby reducing carbon debt from 30 plus year to just one year.

9.5 Closing Remarks

This chapter draws upon the Timilsina and Mevel (2011, 2013) studies to assess climate change mitigation impacts of meeting biofuel mandates and targets introduced by 40 plus countries around the world. International targets set for biofuel expansion require considerable land conversion in order to substitute fossil fuels by volume of consumption. Carbon neutrality will require more than 20 years from 2020, when the announced policies are to be fulfilled and maintained. Notably, this study focuses on first generation biofuels, which require greater land conversion from other activities. Second generation biofuels are not as commercially widespread and have not the robust data for this simulation. The results show that the first generation biofuels will not reduce GHG emissions until 2020 no matter if the new land demand is met from both forest and pasture or only from pasture. The estimates of GHG savings are conservative, since GHG emissions were constrained to CO_2 measurements in this study. Some other GHG savings would occur from rice crops (relatively methane intensive among crops) replacement by biofuel feedstocks (e.g., corn, sugar crops).

References

Al-Riffai, P., Dimaranan, B., & Laborde, D. (2010). *Global trade and environmental impact study of the EU biofuels mandate.* Report for the European Commission, DG TRADE, ATLASS Consortium Institute. Retrieved from http://trade.ec.europa.eu/doclib/docs/2010/march/tradoc_145954.pdf

Dumortier, J., Hayes, D. J., Carriquiry, M., Dong, F., Du, X., Elobeid, A., et al. (2011). Sensitivity of carbon emission estimates from indirect land-use change. *Applied Economic Perspectives and Policy, 33*(3), 428–448.

Fargione, F., Hill, J., Tilman, D., Polasky, S., & Hawthorne, P. (2008). Land clearing and the biofuel carbon debt. *Science, 2008*(319), 1235–1238.

Fischer, G., Hizsnyik, E., Prieler, S., Shah, M., & van Velthuizen, H. (2009). *Biofuels and Food Security, OPEC Fund for International Development.* Vienna: Austria.

Hertel, T. W., Tyner, W, & Birur, D. K. (2010). The global impacts of biofuel mandates", *The Energy Journal, 31*(1), 75–100.

IPCC. (2006). 2006 IPCC Guidelines for National Greenhouse Gas Inventories. In H. S. Eggleston, L. Buendia, K. Miwa, T. Ngara, & K. Tanabe (Eds.), *National Greenhouse Gas Inventories Program* (Agricultural, forestry and other land use, Vol. 4). Kanagawa, Japan: IGES.

Khanna, M., & Crago, C. L. (2012). Measuring indirect land use change with biofuels: Implications for policy. *Annual Review of Resource Economics, 4*, 161–26.

Laborde, D. (2011). *Assessing the land use change consequences of European biofuel policies* (Report for the European Commission, ATLASS Consortium). Retrieved from http://trade.ec.europa.eu/doclib/docs/2011/october/tradoc_148289.pdf

Searchinger, R., Heimlich, R. A., Houghton, F., Dong, A., Elobeid, J., Fabiosa, S., et al. (2008). Use of US croplands for biofuels increased greenhouse gases through land-use change. *Science Express, 319*(5867), 1238–124.

Timilsina, G. R., Beghin, J. C., van der Mensbrugghe, D., & Mevel, S. (2010). *The impacts of biofuel targets on land-use change and food supply: A global CGE assessment* (Policy Research Working Paper 5513). Washington, DC: The World Bank.

Timilsina, G. R., & Mevel, S. (2011). *Biofuels and climate change mitigation: A CGE analysis incorporating land-use change* (Policy Research Working Paper 5672). Washington, DC: The World Bank.

Timilsina, G. R., & Mevel, S. (2013). Biofuels and climate change mitigation: A CGE analysis incorporating land-use change. *Environmental & Resource Economics, 55*, 1–19.

Chapter 10
Biofuel Policies: Subsidy vs. Carbon Tax

Govinda R. Timilsina, Stefan Csordás, and Simon Mevel

10.1 Introduction

Large-scale deployment of cleaner energy sources and technologies through targeted policies is a key measure to reduce GHG emissions. Biofuels are such cleaner sources of energy particularly for the transport sector, a sector that offers only limited opportunities to reduce emissions compared to other sectors such as power and industry. As of yet, biofuel penetration into the transportation fuel mix has only reached about 2 % globally. Biofuels lag in substituting fossil fuels due to high investment costs and rising input prices due to competing demands for feedstock. It is important to note, however, that fossil fuels are cheaper partly because since the pollution and negative externalities are not captured in their prices. Hence, in order to create a level playing field between biofuels and fossil fuels (gasoline and diesel) one needs to subsidize the former or tax the latter at promoting.

This chapter (based on Timilsina et al. 2011) compares two policies to promote biofuels: a direct subsidy to biofuels and a carbon tax to fossil fuels to promote biofuels. Timilsina et al. (2011) use a multi-country, multi-sector recursive dynamic CGE model to assess biofuel penetration and economic impacts of a direct biofuel subsidy and a carbon tax on fossil fuels. They also investigate an option where a carbon tax is imposed on fossil fuels and part of the tax revenue is used to subsidize biofuels. For more details on the model and an in-depth analysis of their results, please refer to Timilsina et al. (2011).

G.R. Timilsina (✉)
The World Bank, 1818 H Street, N.W., Washington, DC 20433, USA
e-mail: gtimilsina@worldbank.org

S. Csordás • S. Mevel
United Nations Economic Commission for Africa, P.O. Box 3001, Addis Ababa, Ethiopia

G.R. Timilsina and D. Zilberman (eds.), *The Impacts of Biofuels on the Economy, Environment, and Poverty*, Natural Resource Management and Policy 41,
DOI 10.1007/978-1-4939-0518-8_10,

10.2 Simulation Scenarios

A range of global, uniform carbon taxes was simulated: from US\$10/tCO_2 to US\$ 100/tCO_2. Per simulation, carbon taxes were introduced in 2012 and fixed throughout the study period (through 2020). Three different policy scenarios were modeled:

1. *Baseline*: existing conditions are assumed to continue. Neither a carbon tax nor additional biofuel subsidies are considered. Any potential expansion of biofuels is due to the policies implemented before 2009. The baseline biofuel penetration reaches around 5.5 % by 2020.
2. *Carbon tax alone*: all fossil fuels are taxed based on their carbon content; the entire carbon tax revenue is rebated to households in order to keep government revenues neutral.
3. *Carbon tax with biofuel subsidy*: carbon taxes on fossil fuels as in Scenario 2 above, but the government uses part of the tax revenues to subsidize biofuel development, and transfers the rest of the carbon tax revenue to households as a lump-sum rebate.

10.3 Simulation Results

10.3.1 Comparing Across Scenarios

Figure 10.1 compares the global changes in biofuel penetration by 2020 under alternative carbon tax rates and various levels of biofuel subsidies. Two levels of carbon taxes are considered: (1) US\$25/tCO_2 and (2) US\$50/tCO_2. Under each carbon tax rate, four options are considered for subsidizing biofuels: (1) no subsidy, (2) 10 % of biofuel cost is subsidized, (3) 25 % biofuel cost is subsidized and (4) 50 % of biofuel cost is subsidized.

As illustrated in the figure, a carbon tax alone does not cause a noticeable change in biofuel penetration. A US\$25/tCO_2 carbon tax increases penetration of biofuels by 0.2 percentage point (from 5.4 % in the baseline to 5.6 % in the carbon tax case). If the rate of the carbon tax is doubled to US\$50/tCO_2, biofuel penetration increases by 0.1 percentage point from 5.6 % in Ctax 25 to 5.7 % in the Ctax-50 case.

As portions of the tax revenue are used to finance the biofuel industry, market penetration improves pointedly. If part of the revenues from a US\$50/tCO_2 tax were used to subsidize 25 % of biofuel costs, its penetration would rise by 3.6 percentage points (from 5.4 % in baseline to 9 % in Ctax-50 with 25 % subsidy case). If the subsidy rate is increased to 50 % under the same tax rate (i.e., US\$50/tCO_2), biofuel penetration would rise by 10.3 percentage points. Notably, higher subsidization of biofuels—such as 50 %—at any tax rate resulted in significant increases in biofuel penetration. A 50 % subsidy rate was found to be the relevant maximum, since it translated to increases in biofuel penetration that are the higher than needed to reach policy objectives under plausible assumptions.

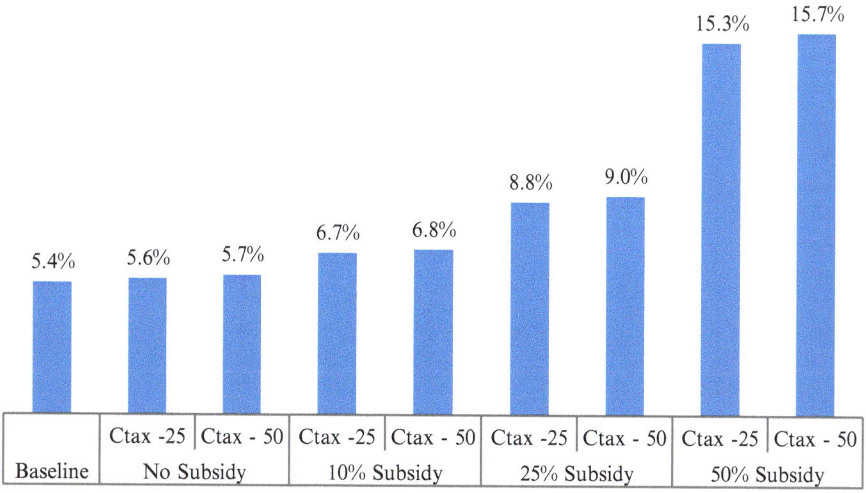

Fig. 10.1 Biofuel penetration at global level with alternative carbon tax and subsidy rates in 2020

10.3.2 Comparing Across Countries, Regions

The effects of different combinations of carbon taxes and biofuel subsidies are compared across countries and regions in Table 10.1. Regionally, as with the world aggregate, a carbon tax alone would do little to stimulate the penetration of biofuels as opposed to a policy of using than tax revenues to subsidize the biofuel industry. For example, a carbon tax of US\$50/tCO_2 would increase biofuel penetration by a maximum 2 percentage points in carbon-intensive economies (e.g., China, Russia). Although less carbon-intensive economies would not experience any significant change in biofuel penetration, higher taxes would depress the economy, reduce total demand for fuels, and, thus, cause the biofuel sector to recede.

A carbon tax in combination with a biofuel subsidy, however, would cause significant increases in biofuel penetration across countries and regions. For instance, a 25 % subsidy would raise most countries' biofuel penetration by over 60 % from the baseline. Subsidies make biofuels seem less costly to produce, compared to the competing products gasoline and diesel. Therefore, landowners would rather commit facilities to biofuel production.

Table 10.2 exhibits the GDP effects of a carbon tax with and without subsidies to biofuels. Given the distortionary effect of a carbon tax, global GDP is reduced under a carbon tax alone. At the US\$25/tCO_2 rate, global GDP recedes by 0.43 % against the baseline scenario. Within this global economic contraction, middle and lower income countries recede relatively more (1.08 %), while higher income nations recede by relatively less (0.07 %). When utilizing part of the tax revenue to subsidize biofuels instead of transfering the entire tax revenue to households, regional

Table 10.1 Biofuels penetration at country/regional level in 2020

Carbon tax	Baseline (%)	US$25/tCO$_2$ (%)			US$50/tCO$_2$ (%)		
Biofuels subsidy		0	25	50	0	25	50
Australia and New Zealand	0.90	0.90	1.50	3.30	0.90	1.50	3.30
Japan	1.10	1.10	1.90	4.00	1.10	1.90	3.90
Canada	2.70	2.80	4.70	9.20	2.90	4.90	9.60
United States	7.70	8.10	12.80	21.90	8.50	13.40	22.70
France	4.40	4.30	7.30	14.50	4.20	7.20	14.20
Germany	5.80	5.80	9.70	18.70	5.70	9.70	18.70
Italy	2.70	2.60	4.60	9.80	2.60	4.50	9.60
Spain	2.30	2.20	3.90	8.30	2.20	3.90	8.30
UK	0.90	0.90	1.50	3.30	0.80	1.50	3.20
Rest of EU and EFTA	1.60	1.60	2.80	6.10	1.60	2.80	6.10
China	5.00	6.20	9.90	17.50	7.00	11.10	19.40
Indonesia	4.20	4.60	7.20	12.80	5.00	7.80	13.70
Malaysia	6.10	6.40	10.70	20.50	6.70	11.10	21.30
Thailand	3.30	3.30	5.10	8.70	3.30	5.10	8.70
Rest of EAP	1.00	1.00	1.70	3.50	1.00	1.80	3.60
India	6.30	6.30	10.00	17.30	6.30	10.00	17.40
Rest of SA	0.90	1.00	1.70	3.70	1.10	1.90	4.10
Argentina	3.80	3.90	6.60	13.10	3.90	6.70	13.20
Brazil	43.00	42.70	53.30	65.80	42.50	53.30	65.90
Rest of LAC	2.30	2.40	4.10	8.10	2.50	4.20	8.40
Russia	7.10	8.20	13.40	24.50	9.10	14.80	26.60
Rest of ECA	2.30	2.40	4.30	8.90	2.60	4.50	9.30
MENA	0.20	0.20	0.40	0.80	0.20	0.40	0.80
South Africa	5.60	6.10	9.40	15.10	6.60	10.00	15.80
Rest of SSA	3.50	3.60	6.00	11.60	3.70	6.20	11.90

economies contract even more. This implies that although using carbon tax revenue to promote biofuels is an attractive policy option from the perspective of biofuel promotion, it may not be necessarily a favorable option from an overall economic perspective.

Breaking down the world aggregates by income levels, middle and lower income nations tend to be affected 3–10 times more than are higher income nations. This reveals a conflict of interest between people at different income levels. Many developing nations (e.g., Brazil, Russia, India, and China) rely on high energy use in manufacturing sectors and tend to rely on coal burning (with the exception of Brazil). The MENA region suffers high GDP loss due to consumption shifts away from oil and gas caused by the carbon taxes.

Subsidies in combination with a carbon tax bear a greater economic burden than a carbon tax alone at every level of taxation. Going from a carbon tax alone (0 % subsidy) to a 50 % subsidization rate, there results an additional 0.1 percentage point loss in world output. Taking the rate of change in GDP with respect to the rate

Table 10.2 Percentage change in real GDP relative to baseline in 2020

Carbon tax	US$25/t$CO_2$			US$50/t$CO_2$		
Biofuels subsidy	0 %	25 %	50 %	0 %	25 %	50 %
World total	−0.43	−0.45	−0.53	−0.85	−0.87	−0.95
High-income	−0.07	−0.08	−0.15	−0.15	−0.16	−0.23
Australia and New Zealand	−0.3	−0.3	−0.31	−0.52	−0.52	−0.53
Japan	−0.02	−0.02	−0.02	−0.06	−0.06	−0.06
Canada	−0.22	−0.23	−0.25	−0.45	−0.46	−0.48
United States	−0.15	−0.17	−0.34	−0.3	−0.32	−0.5
France	0.06	0.07	0.1	0.1	0.11	0.14
Germany	0.07	0.06	0.03	0.09	0.08	0.06
Italy	−0.06	−0.06	−0.07	−0.13	−0.13	−0.15
Spain	0.01	0	−0.02	−0.02	−0.02	−0.04
UK	0.04	0.04	0.04	0.05	0.05	0.05
Rest of EU and EFTA	−0.03	−0.03	−0.05	−0.1	−0.1	−0.12
Middle and low-income	−1.08	−1.12	−1.21	−2.09	−2.14	−2.24
China	−2.1	−2.14	−2.25	−3.87	−3.92	−4.04
Indonesia	−0.45	−0.45	−0.45	−0.94	−0.95	−0.95
Malaysia	−0.35	−0.41	−0.51	−0.89	−0.95	−1.06
Thailand	−0.08	−0.09	−0.11	−0.3	−0.3	−0.32
Rest of EAP	−0.06	−0.06	−0.08	−0.21	−0.21	−0.24
India	−0.86	−0.91	−1	−1.65	−1.69	−1.79
Rest of SA	−0.64	−0.64	−0.67	−1.19	−1.2	−1.23
Argentina	−0.91	−0.91	−0.89	−1.74	−1.75	−1.72
Brazil	−0.16	−0.37	−0.59	−0.35	−0.56	−0.78
Rest of LAC	−0.5	−0.55	−0.66	−0.98	−1.03	−1.14
Russia	−1.96	−2.03	−2.1	−4.17	−4.25	−4.33
Rest of ECA	−0.72	−0.7	−0.67	−1.45	−1.43	−1.4
MENA	−1.23	−1.33	−1.5	−2.57	−2.68	−2.86
South Africa	−0.87	−0.92	−0.98	−1.59	−1.64	−1.71
Rest of SSA	−0.79	−0.84	−0.92	−1.52	−1.58	−1.67

of change in biofuel penetration ("GDP elasticity of biofuel penetration"), the GDP elasticity is higher with a carbon tax alone and lower with subsidies. This makes sense as the greater the number of substitutable goods, the more elastic a product is. In the absence of subsidies no product is favored by non-market forces, so that there is a higher rate of substitutability between products.

Since biofuels are relatively expensive compared to fossil fuels any subsidy level will help the cost-structure of producing firms but muffle firms' incentives to drive costs down themselves. In the near future of 2020, GDP losses ensue as consumers and taxed carbon emitters bear the burden. Thus, carbon taxes with subsidy produce the intended result of deeper biofuel penetration into the energy market; yet, subsidies also distort the market causing deadweight loss and reducing producers' incentives to drive down production costs.

10.4 Impacts on CO_2 Emissions

As expected, the carbon tax is very effective in curbing CO_2 emissions. Even a moderate tax rate of US$10/t$CO_2$ leads to 13 % reduction of global CO_2 emissions from the baseline scenario by 2020 (Table 10.3). A relatively high tax rate of US$50/t$CO_2$ leads to nearly 33 % reduction of CO_2 from the baseline scenario. Comparing across regions of the world, a carbon tax seems to reduce CO_2 at a relatively higher rate in middle and lower income regions than in higher income nations. This is because lower and middle income countries rely on more carbon-intensive energy sources in general, and would experience relatively more emission reductions. For instance, France draws over 80 % of its power from nuclear plants. So even a carbon tax rate of US$50/t$CO_2$ would reduce France's emissions by only 6.5 %. In contrast, the UK (which relies on coal and gas for the bulk of its power generation) would reduce emissions by 28 % under the same tax rate.

Table 10.3 Change in CO_2 emissions (%) relative to baseline in 2020

Carbon tax	US$10/t$CO_2$		US$50/t$CO_2$	
Biofuels subsidy	0 %	25 %	0 %	25 %
World total	−12.9	−13	−32.8	−32.9
High-income	−6.9	−7	−20.6	−20.7
Australia and New Zealand	−10.9	−10.9	−27.2	−27.2
Japan	−6	−5.9	−12.6	−12.5
Canada	−6.7	−6.8	−20	−20.1
United States	−7.8	−8	−23.8	−23.9
France	−1.7	−1.8	−6.5	−6.6
Germany	−4.2	−4.4	−14.4	−14.6
Italy	−2.5	−2.5	−9.6	−9.6
Spain	−2.6	−2.6	−9.9	−9.9
UK	−9.7	−9.7	−28	−28
Rest of EU and EFTA	−5.6	−5.6	−18.2	−18.2
Middle and low-income	−15.8	−15.9	−38.6	−38.8
China	−22.2	−22.4	−50.3	−50.5
Indonesia	−9.3	−9.5	−26.2	−26.4
Malaysia	−8.6	−8.9	−26.6	−26.8
Thailand	−4.5	−4.6	−15	−15.1
Rest of EAP	−8.2	−8.1	−24.2	−24.1
India	−17	−17.2	−38.5	−38.6
Rest of SA	−11.4	−11.5	−31.9	−31.9
Argentina	−15.9	−16	−38.2	−38.2
Brazil	−2.9	−4.1	−10.5	−11.7
Rest of LAC	−8	−8.1	−19.4	−19.5
Russia	−12.6	−12.9	−36.1	−36.4
Rest of ECA	−10.8	−10.8	−30.9	−31
MENA	−8	−8	−26.7	−26.8
South Africa	−21.4	−21.7	−46.3	−46.5
Rest of SSA	−9.9	−10.2	−21.2	−21.4

Interestingly, the rate of carbon emissions reduction is not very high as subsidy rates are raised. The incremental gain from a US\$10/tCO_2 tax alone to 25 % subsidy is only a 0.1 percentage point change in emissions; even a 50 % subsidy produces only a 0.4 percentage point change in world emissions (not shown in the table). This indicates that biofuel subsidies are not effective instruments for GHG emission reductions.

A carbon tax alone reduces demand for the entire fossil fuel group. Subsidies only reduce demand for gasoline and diesel. Therefore, subsidies offer only incremental reductions in carbon emissions by supporting a slightly cleaner fuel. For countries with a larger transportation sector and a relatively cleaner electricity supply system, the effects of subsidies would be higher compared to countries with smaller transportation sectors and more GHG-intensive energy mixes.

10.5 Conclusions

There is ongoing debate on how to promote large-scale deployment of clean energy sources and technologies. Many of the existing literature argue that a carbon tax on fossil fuels would produce a levelized field for clean energy sources to compete. However, the results of Timilsina et al. (2011) show a different picture. A carbon tax does not necessarily cause a large-scale substitution of fossil fuels with clean energy sources. Instead a direct subsidy to clean energy sources would be more effective to increase their shares in the energy supply mix. The study further shows that a carbon tax cum clean energy subsidy scheme where part of carbon tax revenue is used to subsidize clean energy sources would be instrumental to reduce carbon emissions along with promoting the deployment of clean energy sources. This finding largely concurs with Weber et al. (2005) and Barker et al. (2008). The latter two studies had noted that carbon tax impacts on the environment are stronger if carbon tax revenues are used to finance GHG mitigation activities. Yet, using tax revenues to further subsidize carbon mitigating activities would distort market allocations, cause deadweight loss and lead to output contraction. Hence, a cautionary note is in order. While a carbon tax in combination with a clean energy subsidy appears helpful to promote clean energy penetration, taxing one product and subsidizing another creates two distortions causing deadweight loss in the market.

References

Barker, T., Scrieciu, S. S., & Foxon, T. (2008). Achieving the G8 50% target: modelling induced and accelerated technological change using the macroeconometric model E3MG. *Climate Policy, 8*, 30–45.

Timilsina, G. R., Csordás, S., & Mevel, S. (2011). When does a carbon tax on fossil fuels stimulate biofuels? *Ecological Economics, 70*, 2400–2415.

Weber, M., Barth, V., & Hasselmann, K. (2005). A multi-actor dynamic integrated assessment model (MADIAM) of induced technological change and sustainable economic growth. *Ecological Economics, 54*(2–3), 306–327.

Chapter 11
Political Economy of Biofuels

David Zilberman, Scott Kaplan, Gal Hochman, and Deepak Rajagopal

11.1 Introduction

While timber and other biomass have been the main sources of fuel for millennia, there has been an increasing emphasis on growing crops and converting feedstock to liquid fuels (Rajagopal et al. 2009) or for use in power plants. These new fuels were induced by government policies and often require a diversion of resources from agricultural to energy production. Analyzing the performance of biofuels and biofuel policies requires a political economic lens—this chapter will provide such a framework to assess biofuels.

Biofuel policies are significant in that they affect several sectors and policy regimes that have had relatively low levels of linkage in the past. They include agricultural policies, environmental policies, energy policies, and trade policies. Models of political economy can explain what determines the final outcomes that result from these policies as well as the main players that affect the generation of these policies. In order to better understand these effects, we first introduce a background on the political economy literature, and then literature on the political economy of

D. Zilberman (✉)
Department of Agricultural and Resource Economics, College of Natural Resources,
University of California, Berkeley, Giannini Hall 206, Berkeley, CA 94720-3310, USA
e-mail: zilber11@berkeley.edu

S. Kaplan
Department of Agricultural, Resource Economics at the University of California,
Berkeley, CA, USA

G. Hochman
Department of Agriculture, Food, and Resource Economics, Rutgers University,
New Brunswick, NJ, USA

D. Rajagopal
Institute of Environment, University of California Los-Angeles, La Kretz Hall, Suite 300,
Los Angeles, CA 90095, USA

G.R. Timilsina and D. Zilberman (eds.), *The Impacts of Biofuels on the Economy, Environment, and Poverty*, Natural Resource Management and Policy 41,
DOI 10.1007/978-1-4939-0518-8_11,

the agricultural, environmental, and energy sectors. We will then present a conceptual framework, which will be followed with a short discussion of the political economy of policies in various countries (US, EU, Brazil, China, and India).

The range of literature on political economy attempts to explain how political processes are used to allocate resources. A major body of literature emphasizes voting and public choice (Mueller et al. 2008). One basic model focused on in this literature is the median voter model, which explains the need to establish policy packages that are able to garner the majority of support. Another body of literature emphasizes the role of interest groups in affecting regulators in the executive branch. Some of this early literature addresses the regulatory process and Posner (1974) introduced the notion of capture to suggest that regulators' choices may serve the interests of those in regulated industries. Krueger (1974) suggested that the political process may be manipulated through rent-seeking by interest groups and policy makers. Becker (1989) and Peltzman (1976) develop micro models to explain decisions made by policy makers and interest groups, and their implications. The modeling framework developed by Grossman and Helpman (2002) was initially used to explain trade policy regimes and was then expanded to other areas of economic policy. This framework suggests that changes in policy affect the welfare of various groups, yet the weight each group carries in shaping policies differs. Understanding this weight is a major topic of research among political scientists as well as economists such as Olson, who developed frameworks that further explain the formation of interest groups and the factors that determine their relative influence. Rausser et al. (2011) overview a wide range of modeling approaches to assess weights of different groups with emphasis on natural resources and economics. Many of their models assume that policies can be approximated as the outcome of implicit optimization, where policy makers maximize a weighted sum of the welfares of various groups. Furthermore, by observing actual policy choices and estimating the welfare associated with these policies, this approach allows one to assign different weights in the creation of certain policies. We will take such an approach in this chapter as we attempt to explain the determinations of biofuel policies.

The introduction of biofuel will integrate the political economy of agricultural, environmental, and energy policies. There is a large literature on agricultural policy. Schultz (1964) and Cochrane (1979) have suggested that modern agriculture is characterized by a low elasticity of demand for agricultural commodities, high rates of technological innovation, and vulnerability to shocks. This resulted in "the agricultural problem," namely persistent low prices and income. Since much of the increase in agricultural supply originated in public research, government policies in developed countries complimented research with agricultural support policies that aimed to either suppress or stabilize supply and raise income (Rausser et al. 2011). However, in developing countries, farmers were taxed for exports of agricultural commodities, as the government relied on this tax as a major source of revenue. Towards the end of the twentieth century, the number of farmers in developed countries, like the US, declined, and those remaining became relatively well off. Furthermore, the economic growth in Asia has led to an increase in the demand for food, contributing to rising food prices. Thus, agricultural support policies in developed countries have been gradually shifting towards achieving environmental objectives while those in developing countries continue to focus on consumer and producer welfare. Despite periods of higher income and food prices, farmers are still concerned about possible low prices and are looking to expand the range of goods they produce in order to increase demand and raise income, which has led to investment in biofuel.

Political economic considerations have been paramount in the literature on environmental economics. Buchanan and Tullock (1975) argued that when comparing policies of similar efficiency, policy makers may prefer subsidization and direct control over taxes because of the power of regulated industry. Hochman and Zilberman (1978) suggested that intra-distributional consideration within sectors may lead to selection of policies like pollution intensity standards (upper bound on pollution per unit of output) rather than taxes. The importance in terms of political economic consideration has been emphasized in the literature on climate change. While a global carbon tax is clearly an efficient policy, the design of actual policies is heavily affected by distributional considerations. The Kyoto Protocol, for example, treats developed and developing nations differently. It actually includes policy tools (the Clean Development Mechanism) where developed nations subsidize GHGE reductions in developing nations. The recent failure to sign a global agreement has resulted in sectoral and regional policies to address various aspects of climate change, and some of the initiatives that promote biofuel fit within this pattern (McKibbin and Wilcoxen 2002).

Political considerations have been paramount in understanding the economics of the transportation fuel sector. The oil sector has been one of the most dynamic industries in the modern era. Gaining access to cheap oil has become a major policy objective of governments, and oil companies have become major economic institutions. In the 1970s, oil producing countries that form OPEC became a dominant force in determining oil prices globally (Yergin 1991). Yet, some of the models of the oil market use the dynamic Hotelling model of exhaustible resources, expecting prices to rise constantly. However, given the discovery of new oil sources that have led to a steady level and even growth of oil reserves, prices generally declined. Much of the literature on the energy sector consists of static analysis assuming competition. There are several models that consider OPEC as a monopoly. Hochman and Zilberman (2011) find that OPEC takes advantage of its monopolistic power to discriminate between domestic consumers and consumers in importing countries. Political economic considerations affect differences in the prices of energy within OPEC countries, but these prices tend to be significantly lower than in oil importing countries. Hochman et al. (2011a, b) suggest that biofuel consideration will affect both internal and external OPEC prices. We intend to incorporate this framework in our analysis of the political economy of biofuels.

As the transportation sector continued to face increasing demand for fuel, biofuels were introduced as a possible alternative. However, biofuel was originally introduced in the United States around the turn of the twenty-first century as a food additive, and was then expanded to provide alternative fuel. Zilberman et al. (2012) documented the distributional implications of the food vs. fuel tradeoff associated with biofuel production. They suggested that increased supply of feedstock through improved productivity (perhaps GMOs) could counter the reallocation of supply from food to fuel to keep prices low. Zilberman et al. (2012) as well as Chap. 3 on "Biofuels and the Global Food Crisis" suggest that the impact of biofuel policies on food prices, and thus on the welfare of various groups, depends on inventory situations, and periods of low inventory see much higher impacts.

Since biofuels were unable to support themselves early on, government policies were necessary in providing incentives for their production. But these policies have distributional implications. From a welfare economic perspective, de Gorter and Just (2010) and Moschini et al. (2010) suggest that introducing biofuels through mandates is preferred to introducing them through subsidies. Mandates are also preferable from a government perspective, but they harm consumers. However, Tyner and Taheripour (2008) suggested that flexible policies that adjust to both food and fuel situations are ideal. For example, when food prices are high, governments would reduce incentive to reallocate grain from food to fuel, either through a reduction in the specified mandate or a transfer of subsidies from biofuels to food production. This has actually been integrated into current US policy, but the US mandates haven't been waived despite recent years of high food prices, like 2012. Babcock (2012) argued that the introduction of biofuel led to modification of the fuel supply network, and even when the mandates are waived, the amount of grain reallocated from fuel production to food is minimal.

Since one of the objectives of biofuel policies has been to reduce greenhouse gas emissions (GHG), governments introduce various policies to control GHGE of biofuels that are entitled to government support. Two types of policies are renewable fuels standards, which set an upper bound on GHGE of renewable fuels that are required to meet biofuel mandates, and low carbon fuel standards (LCFSs) that set an upper bound on the carbon content of fuel and credit renewable fuels that are used to meet the standard according to their estimated GHGE. Several studies (Chen et al. 2011; Huang et al. 2012; Rajagopal et al. 2012; Drabik and de Gorter 2011; Thompson et al. 2011) demonstrate that the selection of policy matters and different biofuel policies that may reach the same target of GHGE reduction have significantly varying impacts on other policy objectives, and thus the well being of various groups.

The political economy literature, the literature on policies affecting the sectors related to biofuel, and the previous chapters of this book provide a background for a better understanding of the political economy of biofuels in various countries.

11.2 A Conceptual Framework

In the Appendix, we introduce a generalized conceptual framework to understand the basic elements that will determine biofuel policies domestically. The framework consists of six elements:

1. *Interest groups* that may include various groups of consumers, producers, or different government agencies.
2. *Indicators of well-being*, which are variables that affect the welfare of various groups. For example, one indicator of well-being is a change in the price of corn

Table 11.1 Interest groups and policy preferences

	Food consumers	Fuel consumers	Biofuel producers	Taxpayers	Fossil fuel exporting countries (OPEC)	Holders of domestic currency
Mandate	–	(+,–)	+	0	–	+
Biofuel subsidy	–	+	+	–	–	–
GHG tax	–	–	(+,–)	+	–	+
Biofuel tariff	+	–	+	+	+	+
Renewable fuel standard	–	–	(+,–)	0	–	+
Low carbon fuel standard	–	–	(+,–)	0	–	+
Investment tax credit	–	+	+	–	–	+

Note: (+,–) means that the policy may or may not be supported by the given interest group; it is ambiguous

relative to a benchmark. Increases in the price of corn positively affect producers and negatively impact consumers.

3. *Policy parameters* that affect these indicators through market forces and other constraints.
4. *Constraints* that need to be taken into account in determining optimal policy parameters. These constraints include market clearing relationships as well as biophysical, technological, behavioral, and financial limitations.
5. *Preferences* of each group over key indicators that impact them. However, for simplicity, we assume that these preferences translate into weights placed on certain policy parameters.
6. *Political weight* of each group that may reflect its size, coherence, ability to navigate the political system, etc.

We assume that policy variables in a political economic system are determined so as to maximize the weighted sum of the well-being of different interest groups subject to political feasibility constraints. With this optimization problem, policies will determine where the weighted sum of the net marginal benefits to different groups is equal to the weighted sum of costs imposed by the various constraints. This optimization suggests that policies that are desirable to groups with large political weight and are not heavily restricted by constraints are likely to be pursued further than other policies.

Table 11.1 shows several groups we consider important in policy debate about biofuel as well as several policy instruments important with respect to biofuel, and qualitatively hypothesizes how different groups will evaluate each of these policies. In some cases, the directional impact of a policy change on the welfare of a group is apparent. In other cases, it may be ambiguous. For example, with a biofuel mandate, taking away food from food consumers will carry an unambiguous negative effect. On the other hand, a mandate may increase or decrease fuel prices, depending on the marginal cost of biofuel versus gasoline (Zilberman et al. 2011).

Fossil fuel exporting countries do not benefit from biofuel policies under most conditions (Hochman et al. 2011a, b). Holders of domestic assets are the ones that benefit from a high value of the local currency and are concerned with balance of trade deficits. Environmentalists' attitude towards biofuels depends on the net GHGE and other environmental impacts they may generate. Environmentalists may be opposed to biofuel that increases overall GHGs (see Chap. 9) or to biofuels, like palm oil, that are grown in locations that have adverse effects on wildlife. Environmentalists tend to support second-generation biofuels because they are supposed to be "greener" and emit less carbon.

Domestic decisions over biofuel regulation will be affected by the combined attitudes of different groups as well as their relative power. For example, high biofuel mandates that rely on corn ethanol are more likely if the agricultural groups have strong political power and can form a coalition with groups that are concerned with fuel security and exchange rates in order to overcome objections from food consumers and environmentalists. While agricultural industries are relatively small in terms of members, they are politically strong compared to consumer groups because they are well organized and each member invests much more because the marginal benefit to each individual is significant. Combining mandates with biofuel subsidies will be feasible when budgetary constraints are not as limiting. Environmental groups may form a coalition with farmers to establish a LCFS. They also may congregate to push for mandates for second-generation biofuels.

11.3 Special Cases

The conceptual framework presented can help us to look at biofuel policy situations in different countries. Depending on the relative weights of different interest groups in different countries, certain policies may be favored over others. Currently, domestic markets play a much more significant role in biofuel policy determination than global energy and food markets as approximately 90 % of biofuel is consumed domestically (Dufey 2007). We will consider special cases of countries who are significant players in food and fuel markets and place relatively heavy political weight on these two sectors.

11.3.1 United States

In the case of the United States, the mandate and subsidies to corn ethanol reflect the power of the corn sector as well as concerns over food security and balance of trade. The effect on exporters of biofuel is that they do not receive the subsidy given for domestic consumption. The US mandate can be waived when the price of corn is high, but that has not been done, again reflecting the power of the corn and biofuel lobby relative to producers of food and feed. However, continued rises in agricultural commodity prices and resulting pressure from domestic feed consumers as well as food importing nations may lead to relaxation of mandates in order to accommodate food situations. The concerns over budget have led to elimination of the biofuel subsidy. The concern about climate change and the power of environmentalist groups is represented by the implementation of the Renewable Fuel Standard II, the use of indirect land-use effects in computing the GHGE of biofuel, as well as the subsidization of second-generation biofuels. However, the subsidization of second-generation biofuel also reflects concerns over fuel security and balance of trade. The increased abundance of natural gas in the US and the possibility of converting natural gas to fuel for transportation may weaken the concerns about balance of trade and fuel security, which may eventually affect support for biofuel.

In California and several other states, there has been a large emphasis on enacting a Low Carbon Fuel Standard (LCFS). On the surface, this type of standard is superior to a RFS from an environmental perspective, and it reflects the political weight carried by environmentalists. However, as Rajagopal et al. (2012) suggest, when there are LCFSs delegated to only a few states the energy system will adjust, minimizing the impact on GHGE.

11.3.2 Brazil

Brazil established biofuel production because of balance of trade consideration (Goldemberg 2008). In order to overcome the blend limitation, they introduced flex fuel cars so as to increase domestic consumption of ethanol. However, they only utilize 9 million out of at least 30 million feasible acres of agricultural land for biofuel production. Gasoline is taxed quite heavily in Brazil, mostly by the state. On the other hand, biofuel is taxed less heavily than gasoline, and farmers may get subsidized credit for loans to develop land for biofuel production.

Brazil has a national oil company, Petrobras, which is the dominant retailer of gasoline in the country and has a near de facto monopoly on production of gasoline. Petrobras was privatized a little over a decade ago, but the government continues to own a significantly large share and is currently seeking to engage reverse privatization. It seems that Petrobras has significant political power, and that may affect the existing biofuel policies in Brazil (Losekann et al. 2011).

We will provide qualitative analysis of some of the considerations that affect the evolution of biofuel in Brazil. Petrobras has emphasized investment in development of offshore oil exploration (pre-salt) relative to biofuel (Losekann et al. 2011). One possible explanation is that they de facto own the oil reserves that produce the fossil fuel and pay relatively modest royalty while in the case of biofuels, they need to pay farmers for the feedstock and much of the value of the fuel will accrue to the land rent. Furthermore, with the price control and high taxation on fossil fuel in Brazil, Petrobras is on average likely to earn less per unit of fuel from domestic sales than from export. Therefore, Petrobras would prefer to invest in fossil fuel expansion and export as much of the fuel as possible while others will produce biofuel for domestic consumption and Petrobras will earn income as a retailer. On the other hand, companies that do not have access to oil reserves are more likely to invest in development of biofuels. Thus it seems that other entities are more likely to invest in biofuel production in Brazil. They may include domestic companies and multinational oil companies (like Shell, BP, or Total) who look to expand the supply of fuel resources they control.

However, the expansion of biofuel in Brazil requires massive development of regions like Goias, which is ideally located far from the forest but lacks basic infrastructure and manpower, and thus may require imported capital, labor, and transfer of future wealth to new players. Capital from domestic and foreign investors has enabled new technologies and production standards to be implemented. More importantly though has been increased accessibility to foreign markets with large trade barriers because of openness to foreign investment (Moraes 2011).

Much of the foreign direct investment (FDI) since the 1990s has been concentrated in rural sectors because of vast expanses of fertile land, productive crop varieties, the growing demand for food and fuel, and the strength of Brazil's economy (de Andrade and Miccolis 2011). Accelerated investment in biofuel in Brazil in regions where it does not affect deforestation is likely to be optimal from a global perspective, but constraints and uncertainty about foreign land and asset ownership as well as environmental regulation due to political economic considerations will result in a more moderate rate of growth of the biofuel industry (Collins and Erickson 2012; Corbera et al. 2011).

11.3.3 Europe

Europe is highly dependent on the import of fossil fuel from OPEC and other sources, and therefore concerns over fuel security, and in some cases balance of trade, are paramount. The farm lobby in Europe is also very strong, which has led to support for biofuel initiatives. Since there is a much heavier reliability on diesel, the emphasis has been turned to biodiesel production and expansion. However, the profitability of biofuel and biodiesel in Europe is relatively low, so the economic costs of the industry are substantial. High import tariffs on countries with more efficient and sustainable production, like Brazil, distort competition and deemphasize sustainability in an attempt to further domestic biofuel producers (Robbins 2011). Environmental side effects associated with some biofuels have been a source of concern. For example, palm oil production in Indonesia and Malaysia has been associated with deforestation and other ecological damages. Laborde (2011) found that the indirect land-use effect of soybeans and other crops used for biofuel production in Europe is particularly high. Additionally, most biofuel policies as well as the transportation infrastructure in the EU are targeted towards rapeseed-based biodiesel rather than bioethanol, which is more fertilizer intensive and ecologically destructive (Robbins 2011). Thus, the support from environmental groups to biofuel, and especially biodiesel, is likely to be tentative. Rising food prices in 2012 also raised concerns among consumer groups and development advocates, and thus pressured policy decision makers to reduce or limit biofuel mandates in Europe (Euractiv 2012). Europe exemplifies some of the political exchange between different groups that are shaping biofuel policies. As is the case in the US, there is limited potential for first generation biofuel, and hope that the impacts of second-generation biofuel will be more agreeable in terms of environmental and food security issues.

11.3.4 Asia

Asia can be divided between a few wealthy countries like Japan, Korea, and Taiwan and the majority of which are in different stages of development. Japan and Korea are major importers of fuel as well as food, and are likely to be concerned about fuel security. Yet, they do not have substantial biofuel mandates partly because of availability of land but mostly because of a lack of reliable sources of exported biofuels (Biofuels Digest 2012). There have been early attempts to establish biodiesel mandates based on palm oil imports from Malaysia. Furthermore, there is ongoing research to produce biofuel from waste products and algae. Some private companies in India, South Korea, and Singapore are even attempting to outsource biofuel production to other Asian countries with more suitable biofuel production capabilities (Dauvergne and Neville 2010). Thus, political economic considerations suggest that if there was a sufficient and reliable environmentally sound level of supply of biofuel globally, the well-to-do Asian countries would have established policies that utilize it.

In China, there is concern over both fuel and food security. As a country, they are a major importer of food and fuel. So while there is pressure to develop first and second generation biofuel, the support for the biofuel industry will decline during periods of high food prices. China's rapid growth has not been diffused equally among the various social classes, and the implementation of policies favoring biofuels over food production would only increase this gap, as food purchases make up a much larger part of the poor's income than fuel consumption. However, the impact on certain foods may not be as large due to decreased fuel prices as a result of increased biofuel production, considering that fuel is certainly an important input for the production of many different kinds of crops. Another significant factor in deterring biofuel production in China, despite its increasing demand for energy to supply its massive population, is the inefficiency in its scale of production. Much of the biofuel produced in China comes from inefficient corn, which may become exponentially important in times of rising food prices (Robbins 2011). Environmental concerns seem to play a smaller role in China than in Europe, as valuation of the environment is thought of as increasing with income (Mcconnell 1997).

In India, strengths of environmental groups and the food lobby are likely to curtail efforts to produce biofuel from food crops. Yet, the capacity to produce biofuel from sugarcane profitably has led the farm sector to promote this activity, which is subject to debate (Khanna et al. 2013). There is a large tradeoff between using molasses (a byproduct of sugarcane that can be used to produce ethanol) for alcohol production or biofuel production. According to the model presented by

Khanna et al. (2013), if 50 % of molasses must be used to produce biofuel, food prices will not change very much but prices of alcohol may nearly double. However, a larger mandate will benefit India's balance of trade due to a reduction in gasoline imports, which currently make up over 75 % of petroleum use within the country, and in whole decrease the net trade deficit by about 30 % (Khanna et al. 2013). Thus, there is a near consensus that when possible, the introduction of energy crops like jatropha on marginal land is desirable. Yet its economic feasibility still remains in question. Clearly, the political forces in developing countries are less likely to support switching from grain to biofuel production, but will support introduction of energy crops as new sources of income for the rural sector, especially when they do not compete significantly with utilized resources.

11.4 Conclusions

Political economic considerations are paramount to the evolution of biofuel policies because the recent emergence of the biofuel industry was triggered by such government policies. Development of biofuel is desirable from perspectives of fuel security and improving balance of trade, and is very heavily supported by the agricultural lobby. Yet, it may encounter very strong objections by oil companies, OPEC countries, the feed sector, and food consumers. Environmental groups will support biofuels that reduce GHGE and do not result in deforestation. They are less likely to support corn in the US and biodiesel in Europe, but biofuel in Brazil is supported as long as it is associated with regulations that protect forests from clear cutting.

Different countries will implement different types of policies because of different economic and biophysical conditions as well as differences in the relative weights given to various groups and, as a result, policy objectives. A major factor deterring biofuel expansion in developing countries, despite the fact that many of them are located in ideal geographic locations for biofuel production, is the inability to introduce effective policies and corruption (Robbins 2011). First-generation biofuels produced from food products have been controversial, yet they have been introduced in the US and Europe, and less so in developing countries like China and India. Unlike corn and soybeans, sugarcane is one first-generation biofuel that is far from reaching its potential as a clean substitute for fossil fuel. While Brazil's biofuel sector has grown, the interest of Petrobras in developing oil and the desire to limit international control of the industry has slowed its expansion. There has been strong support for to the expansion of second-generation biofuels throughout the world, especially when it provides opportunities to utilize marginal lands. However, thus far the cost effectiveness of these biofuels has been limited, but there has been significant support to research these technologies and subsidies for their introduction in some places, and with additional knowledge they may become commercially viable and contribute to address both GHGE and fuel scarcity challenges.

Appendix

The framework can be generalized to establish policies internationally. We assume that the country has I interest groups and let $i=1,\ldots,I$ an interest group indicator. The relevant interest groups in the context of biofuels include food consumers, food producers, fuel consumers, biofuel producers, fossil fuel producers, environmentalists, automobile companies, and macro policy makers. Each interest group is assumed to have a political weight, β_i, as well as welfare function $W_i=g_i(X)$ where X is a K dimensional vector of indicators of well-being (the index $k=1,\ldots,K$), and the element of the vector X_i is x_k. These indicators can be both quantities and prices, and they reflect performance relative to a benchmark. They include changes relative to a benchmark in the prices of food and fuel, GHGE, government revenue, balance of trade, and food security (measured by the amount imported from less secure regions), as well as other indicators. Thus, if $k=1$ indicates food price, x_1 is a change in the price of food relative to a benchmark, and $g_1(x_1,\ldots,x_k)$ represents the effect of the changes in indicators that affect the well being of the first interest group.

Each indicator is a function of biofuel policy variables. Let Y be an N dimensional vector of policies $n=1,\ldots,N$, and the nth element of Y is y_n. Each policy indicator is a function of the biofuel policy variables, and let the function $x_k=\varphi_k(Y)$ denote this functional relationship. The policy variables we consider may include the level of biofuel tax, a biofuel mandate, biofuel subsidies, biofuel tariff, GHGE regulations related to biofuel, regulation on the use of biofuel (the blend wall). The relationship between the indicators and policy is determined by market interaction and other constraints that affect the economy, which are not specified here. To simplify our analysis, we assume that there is a direct functional relationship between welfare of each group and policy measures, denoted by $f_i(Y)$ for $i=1,\ldots,I$. Thus, $W_i=g_i(x_1(Y),\ldots,x_k(Y))=f_i(Y)$. With this notation, if $i=1$ is food consumers, $f_1(Y)$ is the effect of a proposed policy y on the well being of consumers (through its effect on the different performance measures).

We will assume like a few of the models overviewed by Rausser et al. (2011) that the determination of a policy parameter can be determined by the maximization of a political economic objective function $S=h(f_1(Y),\beta_1,\ldots,f_I(Y),\beta_I)$, which is assumed to be well-behaved. For example, this objective function can be:

- Linear $S=\sum_{i=1}^{I}\beta_i f_i(Y)$
- Log linear $S=\sum_{i=1}^{I}\beta_i \log f_i(Y)$

The determination of optimal policy is subject to a variety of constraints, including budgetary limits, biophysical limitations on a feasible range of policy parameters, etc. We assume that there are M constraints and the indicator of constraints is $m=1,\ldots,M$, and the mth constraint is $c_m(y_1,\ldots,y_n)=0$ where c_m is a well-behaved function of the policy parameters.

Thus, the political system will result in policy parameters y_n that solve:

$$\max_{y_1,\dots y_n} h\left(f_1\left(Y\right),\beta_1,\dots,f_I\left(Y\right),\beta_I\right) \text{ s.t. } c_m\left(y_1,\dots,y_n\right)=0 \text{ where } m=1,\dots,M$$

The first-order condition that will determine the policy parameter y_i is:

$$\sum_{i=1}^{I}\frac{\partial h}{\partial f_i(Y)}\frac{\partial f_i(Y)}{\partial y_n}-\sum_{m=1}^{M}\lambda_m\frac{\partial c_m}{\partial y_n}=0 \tag{11.1}$$

Where $n=1,\dots N$

where λ_m is the shadow price of the mth constraint. The first-order condition (1) suggests that the optimal level of y_n is where the weighted sum of marginal contributions is equal to the sum of the marginal cost it imposes on the constraints. In the linear scenario, Eq. (11.1) becomes (11.2).

$$\sum_{i=1}^{I}\beta_i\frac{\partial f_i(Y)}{\partial y_n}-\sum_{m=1}^{M}\lambda_m\frac{\partial c_m}{\partial y_n}=0 \tag{11.2}$$

Equation (11.2) suggests that policies that are desired by interest groups with large political weight will gain more support. However, political weight is not the only factor that affects policy determination. It is affected by the impact on marginal benefit to various groups and to what extent the policy is constrained by the factors mentioned above.

References

Babcock, B. A. (2012). *Preliminary assessment of the drought's impacts on crop prices and biofuel production* (CARD Policy Brief 12-PB-7 2012). Ames, IA: Center for Agricultural and Rural Development, Iowa State University.

Becker, G. S. (1989). Political competition among interest groups. In J. F. Shogren (Ed.), *The political economy of government regulation* (pp. 13–27). Dordrecht, The Netherlands: Kluwer Academic.

Biofuels Digest. (2012). *Biofuels mandates around the world.* Retrieved from http://www.biofuels-digest.com/bdigest/2011/07/21/biofuels-mandates-around-the-world/

Buchanan, J., & Tullock, G. (1975). Polluters' profits and political response: Direct controls versus taxes. *American Economic Review, 65*, 139–147.

Chen, X., Huang, H., Khanna, M., & Önal, H. (2011). *Meeting the mandate for biofuels: Implications for land use, food and fuel prices* (NBER Working Paper No. 16697). Cambridge, MA: National Bureau of Economic Research.

Cochrane, W. (1979). *The development ofAmerican agriculture: A historical analysis.* Minneapolis, MN: University of Minnesota Press.

Collins, G. & Erickson, A. (2012, March 28). Tilling foreign soil: New farmland ownership laws force Chinese agriculture investors to shift strategies in Argentina and Brazil. *China SignPost, 57.*

Corbera, E., Estrada, M., May, P., Navarro, G., & Pacheco, P. (2011). Rights to land, forests and carbon in REDD+: Insights from Mexico, Brazil and Costa Rica. *Forests, 2*(1), 301–342.
Dauvergne, P., & Neville, K. J. (2010). Forests, food, and fuel in the tropics: The uneven social and ecological consequences of the emerging political economy of biofuels. *The Journal of Peasant Studies, 37*(4), 631–60.
de Gorter, H., & Just, D. R. (2010). The social costs and benefits of biofuels: The intersection of environmental, energy and agricultural policy. *Applied Economic Perspectives and Policy, 32*(1), 4–32.
de Andrade, R. M. T., & Miccolis, A. (2011). *Policies, institutional and legal framework in the expansion of Brazilian biofuels* (Working Paper). Bogor, Indonesia: CIFOR.
Drabik, D., & de Gorter, H. (2011). Biofuel policies and carbon leakage. *AgBioForum, 14*(3), 104–110.
Dufey, A. (2007). International trade in biofuels: good for development? And good for environment? *Environment for the MDGS: An IIED briefing.*
Euractiv. (2012). *OECD agriculture official urges end to biofuel mandates*. Retrieved from http://www.euractiv.com/cap/oecd-news-514884.
Goldemberg, J. (2008). The Brazilian biofuels industry. *Biotechnology for Biofuels, 1*, 6.
Grossman, G., & Helpman, E. (2002). *Interest groups and trade policy*. Princeton, NJ: Princeton University Press.
Hochman, E., & Zilberman, D. (1978). Examination of environmental policies using production and pollution microparameter distributions. *Econometrica, 46*(4), 739–760.
Hochman, G., Rajagopal, D., & Zilberman, D. (2011a). The effect of biofuels on the international oil market. *Applied Economic Perspectives and Policy, 33*(3), 402 427.
Hochman, G., Rajagopal, D., & Zilberman, D. (2011b). The effect of biofuels on the international oil market. *Applied Economic Perspectives and Policy, 33*(3), 402–427.
Hochman, G., & Zilberman, D. (2011). *The political economy of OPEC* (Working paper). Berkeley, CA: University of California at Berkeley.
Huang, H., Khanna, M., Onal, H., & Chen, X. (2012). Stacking low carbon policies on the renewable fuels standard: Economic and greenhouse gas implications. *Energy Policy, 56*(C), 5–15.
Khanna, M., Önal, H., Crago, C. L., & Mino, K. (2013). Can India meet biofuel policy targets? Implications for food and fuel prices. *American Journal of Agricultural Economics, 95*, 296–302.
Krueger, A. O. (1974). The political economy of the Rent-Seeking Society. *American Economic Review, 64*, 291–303.
Laborde, D. (2011). *Assessing the land use change consequences of European biofuel policies*. Washington, D.C.: International Food and Policy Research Institute.
Losekann, L., Ribeiro, E. P. Bone, R. B., & de Oliveira, O. (2011). *Energy restrictions to growth: The past, present and future of energy supply in Brazil*. Paper prepared for the Ilhabela Conference.
McConnell, K. (1997). Income and the demand for environmental quality. *Environment and Development Economics, 2*, 383–399.
McKibbin, W. J., & Wilcoxen, P. J. (2002). *Climate change policy after Kyoto*. Washington, D.C.: Brookings Institution Press.
Moraes, M. (2011). Perspective: Lessons from Brazil. *Nature, 474*(7352), S25.
Moschini, G., Lapan, H., Cui, J., & Cooper, J. (2010). Assessing the welfare effects of US biofuel policies. *AgBioForum, 13*(4), 370–374.
Mueller, D. C., Rowley, C. K., & Schneider, F. G. (2008). *Readings in public choice and constitutional political economy* (pp. 31–46). New York: Springer.
Peltzman, S. (1976). Toward a more general theory of regulation. *The Journal of Law and Economics, 19*.
Posner, R. (1974). Theories of economic regulation. *Bell Journal of Economics and Management Science, 5*, 335–358.

Rajagopal, D., Sexton, S., Hochman, G., & Zilberman, D. (2009). Recent developments in renewable technologies: R&D investment in advanced biofuels. *Annual Review of Resource Economics, 1*, 621–44.

Rajagopal, D., Zilberman, D., & Hochman, G. (2012). *Multicriteria comparison of fuel policies: Renewable fuel mandate, emission standards, and carbon tax*. Berkeley, CA: University of California (Working Paper).

Rausser, G. C., Swinnen, J., & Zusman, P. (2011). *Political power and economic policy: Theory, analysis, and empirical applications*. New York: Cambridge University Press.

Robbins, M. (2011). Policy: Fuelling politics. *Nature, 474*, S22–S24.

Schultz, T. W. (1964). *Transforming traditional agriculture*. Chicago: University of Chicago Press.

Thompson, W., Whistance, J., & Meyer, S. (2011). Effects of US biofuel policies on us and world petroleum product markets with consequences for greenhouse gas emissions. *Energy Policy, 39*(9), 5509–5518.

Tyner, W., & Taheripour, F. (2008). Biofuels, policy options, and their implications: Analyses using partial and general equilibrium approaches. *Journal of Agricultural & Food Industrial Organization, 6*(2), 1–19.

Yergin, D. (1991). *The prize: The epic quest for oil money and power*. New York: Simon & Schuster.

Zilberman, D., Hochman, G., & Rajagopal, D. (2011). On the inclusion of indirect land use in biofuel. *University of Illinois Law Review, 413*.

Zilberman, D., Gal, H., Deepak, R., Steve, S., & Govinda, T. (2012). The impact of biofuels on commodity food prices: Assessment of findings. *American Journal of Agricultural Economics, 95*, 275–281.

MIX
Papier aus verantwortungsvollen Quellen
Paper from responsible sources
FSC® C105338

If you have any concerns about our products,
you can contact us on
ProductSafety@springernature.com

In case Publisher is established outside the EU,
the EU authorized representative is:
Springer Nature Customer Service Center GmbH
Europaplatz 3, 69115 Heidelberg, Germany

Printed by Libri Plureos GmbH
in Hamburg, Germany